TO ENJOY TO LISTEN

视听盛宴

图书简介

一片胶卷录制着电影的梦幻，一帘幕布掀开戏剧的震撼，一条五线谱歌唱出音乐的美妙，一把小提琴演奏出艺术的魔力——电影院、剧院、音乐厅与演艺中心四个章节共同为读者打造了一场以书为介质的"视听盛宴"。本书项目精美独到，版式清新自然。沉浸其中，读者似乎可于图文之间体验到画面的转换与声音的跃动。除了华丽的彩色照片，书中还插入绘制精细的图纸，附有相关建筑设计规范与权威研究，为读者形象地列举了剧场建筑中观众厅、舞台、门厅、音效等的优秀设计范例。在影视音乐中交换心灵，在《视听盛宴》中感悟设计！

TO ENJOY TO LISTEN is composed of four chapters: cinemas, theaters, music halls and performing arts centers. The book aims not only visually beautiful, but also indeed practical, telling some design skills of theater buildings with a combination of gorgeous photos, elaborate drawings, a fresh layout and vivid descriptions. It even shows in a particular way with images of auditoriums, stages, foyers and corridors etc. Readers may get a deep understanding of architectural design while enjoying plays and music here. Films record the magic of a movie; curtains draw the shock of a play; staffs compose the wonder of a piece of music; violins show the power of art....

所有人名按字母顺序排列　Name in alphabetical order

Publisher 出版人　Sun Xueliang
Distributor 发行人　Sun Xueliang
Committee 编委　Alessandro Corradini, Andre Kikoski, Arseniy Borisenko, Carmelo Tumino, Catherine Holliss, Christian Kienapfel, Cristiano Cosi, David Rockwell, Doriana Fuksas, Dwayne Oyler, Emile van Vugt, Erik Alden, Gaetano Manganello, Giovanni Vaccarini, Jacques Ferrier, Jenny Wu, Jesús MaríaSusperregui Virto, Jorge Gracia, Marcello Marchesini, Massimiliano Fuksas, Misak Terzibasiyan, Nick Barratt-Boys, Paul Gates, Peter Ruge, Peter Zaytsev, Rand Elliott, Reese Rowland, Robert Oshatz, Valerio Barberis, Whitney Sander, Zhenya Merkulova

Editor-in-Chief 执行主编　石莹 Erica
Marketing Manager 市场总监　林佳艺 Rita
Editor-in-Charge 编辑部主任　张晓华 Kitty
English Editors 英文编辑　毛玲玲 Marian　石莹 Cindy
王培娟 Joan
Art Editor 美术编辑　夏金梅 Candy
Contributing Editors 特约编辑　Adam Licht, Ana Román Escobar, Andre Kikoski, Anne Sophie Gauclin, Christian Kienapfel, Courtney DeMesme-Anders, Dwayne Oyler, Elodie Leneveu, Emile van Vugt, Federica Provaroni, Giovanna Lamolinara, Hanneke Buurman, Jorge Gracia, Matthias Matschewski, Michelle Jameson, Misak Terzibasiyan, Paul Gates, Reese Rowland, Robert Oshatz, Sheela Pawar, Stephanie Glen, Susanne Lambert, Vadim Costyrin, Zhenya Merkulova

Price 定价　USD 29　RMB 98
Add: 上海市虹口区天宝路886号天宝商务楼3楼
3rd Floor, NO. 886 Tianbao Road, Hongkou District, Shanghai, China
Post Code: 200086
Tel: 86-21-65018097-378
Fax: 86-21-65018097-188
E-mail: info@internationalnewarchitecture.com
Website: www.internationalnewarchitecture.com

Media Link 媒体支持

Supportive Companies & Associations 理事单位

EADG 泛亚国际

图书在版编目（CIP）数据

国际新建筑. 11 / 凤凰空间·上海编. -- 南京：
江苏人民出版社, 2012.10
ISBN 978-7-214-08789-8

Ⅰ. ①国… Ⅱ. ①凤…Ⅲ. ①建筑设计－作品集－世界－现代 Ⅳ. ①TU206

中国版本图书馆CIP数据核字(2012)第231221号

国际新建筑11
策划编辑：石　莹　林佳艺
责任编辑：蒋卫国
特约编辑：王培娟
出版发行：凤凰出版传媒股份有限公司
　　　　　江苏人民出版社
　　　　　天津凤凰空间文化传媒有限公司
邮　　编：210009
印　　刷：上海当纳利印刷有限公司
开　　本：965毫米×1270毫米　1/16
印　　张：12
字　　数：96千字
版　　次：2012年10月第1版
印　　次：2012年10月第1次印刷
书　　号：ISBN 978-7-214-08789-8
定　　价：98.00元
销售电话：022-87893668
网　　址：www.ifengspace.com
(本书若有印装质量问题，请向销售部调换)

■《国际新建筑》
International New Architecture

卷首语

古往今来，欧美建筑文化已有深厚的历史渊源，其设计风格与内涵亦是博得世人景仰与膜拜。欧美人很重视楼梯，从材质到品质，再到风格，都比较讲究。欧洲人甚至认为，在一栋住宅里，楼梯是主人地位和尊严的象征，也是主人性情和生活品位的具体表现。从建筑艺术和美学的角度看，楼梯是建筑艺术视觉的焦点，楼梯的风格、材质、色调要同周围空间环境以及其他摆设相协调。

近年来，随着城市住宅郊区化趋势的发展，大量独栋别墅、联排别墅与跃层住宅的出现，为楼梯艺术提供了很大的展示空间。本书就目前楼梯的发展趋势，为读者呈现了欧美建筑一系列金属、玻璃、纯钢、实木的单跑或是多跑楼梯，包括一字形、L形、半椭圆形或螺旋形等设计，引领建筑领域时代潮流。这些楼梯制作工艺精良、风格多样化，无论是华贵典雅、简洁素净，还是暗淡仿古，都为空间布局与设计奠定了基本格调。

楼梯已不仅仅是一件单纯的连接各楼层的工具，也不是一件普通的摆设，它凝聚了空间的美感，流露出房屋主人独有的韵味。

European and American architecture has seen a long history through ages. People learn about the design skills and gain admiration for architects there. As is known, most Europeans and Americans pay much attention to the staircase in a building, whether the material selection or design styles. Some people in Europe even take the stairs as a symbol of their social status and dignity; for they believe the stairs show a lot about their personalities and tastes. From another point of view, the staircase is the focal point of a building in architectural aesthetics—the design styles, materials and colors shall match harmoniously with other facilities.

The staircase design develops further and rapidly in recent years, with the invention of single-family house, detached house and duplex house in suburb area. This issue will show some metal, glass, steel and wooden staircases to our lovely readers—each has taken a unique style, luxurious and elegant, clean and modest, refined and classic; each is distinct in form, straight line, L shaped, half-oval or spiraled. They are made delicate and fashionable, settling the interior mood.

The staircase not only functions as an access to connect different levels nowadays, but also a medium to express a specific way of living in a building, an area, a city.

Staircase Design
— 楼梯设计 —

www.internationalnewarchitecture.com

Contents 目录

专题 Feature

006　环环相扣——陈内卡住宅
Chenequa Residence

雕塑般的楼梯好似生长在石柱表面的藤类植物,沿着中庭盘旋而上,暗红色流露出的华贵与砖石原始的素净一唱一和,打造出低调而雅致的空间。
The sculptural stairs, like vines, spiral upwards around the stone column in central atrium. The magnificent dark red treads and handrails make a harmonious contrast with natural bricks, adding an elegant atmosphere.

022　旋风楼梯——阿玛尼第五大道店
Armani Fifth Avenue

旋风般的楼梯蜿蜒盘绕其中,将每一层连接起来,形成了令人震撼的视觉效果。它以辐射状的钢结构为骨骼,塑性材料为皮肤,如巨大的龙卷风从这里"席卷"而过,将每一层的空间连接起来。
The heart of the building is, in fact, epitomized by a power generated dancing vortex—the staircase, structure in rolled calendar steel and clad in a plastic layer that highlights its exceptional sculptural presence.

032　电线楼梯
Live Wire

这个楼梯使用了约732米的铝金属管和金属棒,由各种类型的复合环组成,将必要的楼梯元素结合在了一起。同时,楼梯表面采用了厚度不一的小型穿孔铝板,打造了一个从台阶到护栏,再到天篷的连续半透明表面。
Constructed of approximately 2,400 linear feet of aluminum tubing and rods, the stair employs a combination of complex loops that perform a variety of tasks as they merge together to form the necessary stair elements. Similarly, the stair incorporates faceted perforated aluminum panels of two different thicknesses to create a continuous, semi-transparent surface from stair tread to guardrail to canopy.

042　形散而神聚——错位住宅
Split View

由黄杨木建造的楼梯将错落开来的各层连接成为统一的整体,朴质的色彩奠定了内部空间的基调,突出"形散而神聚"的独特魅力。
The Yellow Poplar stairs unify the "split levels", seeming apart in form but all in one indeed, speaking about interior design in a plain way with a simple color palette.

048　宁静以致远——E3住宅
Maison E3

用钢铁和胡桃木打造的中央楼梯明亮通透,为这座轮廓清晰的E字形住宅增添了缕缕舒适感。
The central stairs, light and airy, are made of steel and walnut, which create a sense of comfort to the E shape house.

056　圣海伦岛斯图尔特博物馆
Stewart Museum at St. Helen's Island

这座楼梯以全透明的外观连通博物馆上下三层,成为这个庭院里闪亮的一角;而电梯则四周被镜面围合,营造出万花筒般的建筑效果,将新建筑和周边浓厚的历史氛围紧密地结合起来。
The architectural transparent glazed staircase connects the circulation through the museum's three levels; while the elevator is covered with mirrors to create the effect of a kaleidoscope which binds the new construction to its historical context.

062　"钢"柔并济——格林尼治村联排式别墅区
Greenwich Village Townhouse

钢质楼梯呈螺旋状上升,直至屋顶,增强了垂直方向的空间感,屋顶半椭圆形的天窗则为楼梯口引入光亮,更显得室内宽敞明亮。
The sense of vertical flow between levels is promoted by the spiraling movement of the half-oval open-riser steel stair terminating in a small skylit room at the roof level.

068　树屋
Tree House

内部楼梯由1.3厘米厚的铝板制成,与双层楼高的暗薰衣草色石板壁炉墙相合,为客厅引入雕塑元素;外部悬臂式楼梯则是观赏下方溪流美景的绝佳地……
The inside stairs are made of 1/2" aluminum plate, which create a sculptural element for the dramatic living room and provide a counterpoint to the double-height fireplace wall that is clad with dark lavender slate. Two exterior cantilevered stairs provide bar grate steps overlooking the stream below.

072　全新一体化办公室
New Sincretica Offices

两部楼梯为办公室内部添色不少。楼梯由金属架支撑,木板台阶看上去好像是悬浮在空中,下置"穿孔金属板"加以支撑。楼梯的栏杆由一根金属管构成,管道构造了一处由一楼通往阁楼的宏伟"线条"。
The interior space is characterized by the double flight of stairs. It consists of a metal frame on which the treads are almost suspended wood, supported by a perforated metal cabinet. The handrail is a metal tube that draws a strong line from the ground up to the loft where it becomes parapet.

建筑 Architecture

076 拍谱海锡克&查理海锡克香槟总部大楼
Head Office for Piper Heidsieck and Charles Heidsieck Champagnes

084 白色箱体——雷乌斯112大楼
112 Building in Reus

096 T别墅
Villa T

104 浑然一体——南加州林间别墅
A Modern Villa

114 ROC社区教育中心
ROC Rijn Ijssel

120 杭州新市政大楼
Hangzhou Congress Center

128 阿肯色州研究院
Arkansas Studies Institute

136 巨石——惠灵顿国际机场客运航站楼
The Rock—Wellington Airport International Passenger Terminal

144 清新自然风——瓜达鲁普阁楼式创意酒店
Endémico Resguardo Silvestre

150 蒙塔尔托迪卡斯特罗新剧院
New Theater in Montalto di Castro

156 瑞士大使馆
Swiss Embassy

室内 Interior

162 MiMA豪华住宅区
MiMA

172 501咖啡馆
Café 501

180 第二家园
Second Home

188 Yandex 喀山办公室
Kazan Yandex Office

Chenequa Residence
环环相扣——陈内卡住宅

| Robert Harvey Oshatz Architect |

旋转楼梯
Spiral stair

陈内卡住宅的设计过程分为两个阶段。第一阶段在原先的建筑里进行，包括了主要的生活区域。而稍晚进行的第二阶段建设则包括了玻璃顶棚的游泳池和为老年人建造的套房。

鉴于地形较为特殊，住宅的入口被设置在这个三层建筑的第二层。在入口层可以清晰地望见湖泊美景，并与客厅、餐厅和厨房等住宅其他空间相通。这些空间用低矮的天花板与大楼中庭隔绝，保证了内部隐秘的环境。房屋平面被四周隆起的地势包围，上下被玻璃覆盖，使壮丽的湖泊景色映入房内各个角落。建筑平面从厨房延伸至在树丛中蜿蜒的露天平台。弯曲的钢扶手围绕着阳台，连接厚重的花园围墙，有效地使漂浮的建筑扎根地面。在房屋的另一端，旋转外墙结构勾勒出围绕高大橡树而建的客厅，用相似的方式使建筑回归地面。

入口层下方的楼层为居家生活而建。主中庭下方有游戏室、小酒吧、小影院和书房。而儿童房和一间客房则建于侧楼，在厨房楼下。公共居家场所则位于地下，室外的景观不时被厚重的石墙遮挡。由于斜坡的缘故，儿童房的地势稍高，占据有利地形，但是也受制于外面的石头并被固定在地面上。底楼的流通区域相

对于高处的公共空间更为封闭。一条高而狭窄的走廊通向卧室,走廊仅用玻璃装饰。作为住宅内少数的非景观空间,走廊使人感到安全和温馨。

与主中庭不同,卧室在大小和形状上更为传统,那里可以看见湖泊,但是只有一个侧影。位于走廊尽头的卧室高度甚佳,确保了它的私密性,露天平台底部是从厨房延伸出的夸张天花板。如果说入口层观景甚佳,四面开阔,那么底楼的居家空间则充满了安全温馨的氛围。

因客户表示不在乎建筑对称性,Oshatz用放射形布局代替了对称结构。住宅平面图由一系列不同圆心的圆面组成。为了完成合理协调的平面规划,每一个圆面都相互交错。主圆面勾勒出建筑轮廓线,沿小路而建,避免移除任何树木。这些圆弧也呼应了湖泊的形状,使房间如同向周围景观敞开怀抱。一棵大橡树成为住宅的中轴。而每个圆面的外形互相呼应,使人们体验到空间的自如与和谐。

从入口处望去,住宅的体积并不大,但是利用陡峭的地势最大化了湖泊的景观和联系。最终的结构如同在地上涌现出的一系列平面,顺着山的水平面绵延。每一处平面都很独特,互不盲从;它们质地轻盈,被玻璃分隔。屋顶与楼面材料不同,有它们独特的图案,使大楼设计复杂多变。轻盈、活力四射的水平面由取自当地的垂直石材支撑,可增强建筑的稳定性。

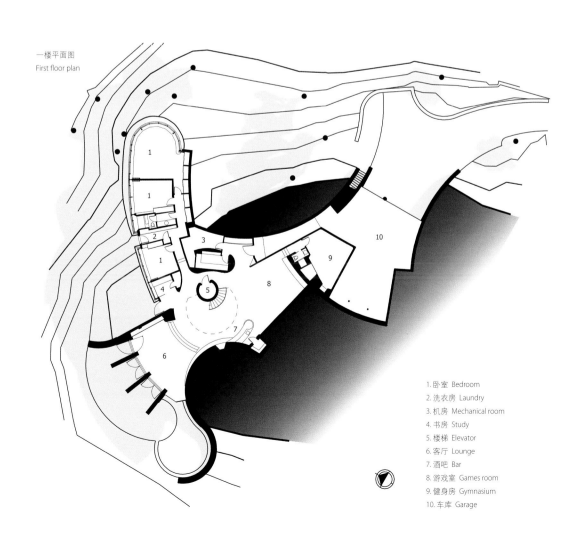

一楼平面图
First floor plan

1. 卧室 Bedroom
2. 洗衣房 Laundry
3. 机房 Mechanical room
4. 书房 Study
5. 楼梯 Elevator
6. 客厅 Lounge
7. 酒吧 Bar
8. 游戏室 Games room
9. 健身房 Gymnasium
10. 车库 Garage

中央楼梯
Atrium staircase

向上看
Looking up

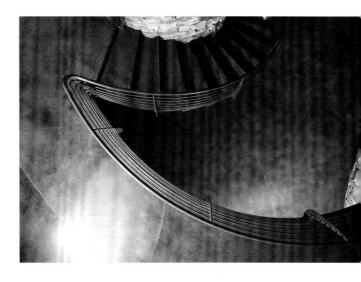

The Chenqeua Residence has been designed to be constructed in two phases. The extant building constitutes the first phase of construction and includes the primary living areas for the family. The second phase of construction to be undertaken at a later point will consist of a glass roofed swimming pool area and a suite for visiting grandparents.
By building into the site, entrance to the house is made on the second of three floors. The entry level is provided with unobstructed views to the lake and accommodates the public spaces of the building including the lounge room, dining room and kitchen. These spaces are separated from the main atrium by low ceilings that help to provide an intimate environment. The plan of the house wraps itself around the convex topography of the site which, when combined with the use of floor to ceiling glass, ensures that magnificent lake views are seen from all the internal spaces. The floor plan is continued out through the kitchen and onto a cantilevered deck that extends out amongst the trees. A ribbon of curved steel balustrades bends itself around the balcony and returns

主浴室
Master bathroom

厨房
Kitchen

厨房走廊
Kitchen Porch

into a heavy stone garden wall, effectively tying the floating floor plain back to the earth. On the other side of the house, the helical stone wall that defines the lounge room twists around a tall oak tree and ties the floor back to the ground in a similar fashion.

The floor below the entry level is designed for family-based functions. Below the main atrium space is a games room, a small bar, a theatre room and a small study. In the wing that follows under the kitchen are the children's bedrooms and a guest room. The communal family spaces are dug below the ground and the view is always filtered by heavy stone walls where they open up to the exterior. The children's bedrooms are afforded elevation and views by virtue of the sloping site but are bound by stone and anchored to the ground. The circulation on the bottom floor is also much more guarded than that provided to the public spaces above. The bedrooms are accessed via a tall but thin corridor adorned only with highlight windows. As one of the few spaces in the house without commanding views, the corridor provides a sense of security and warmth.

Unlike the main atrium space, the bedrooms are more conventionally scaled and shaped. They provide views to the lake, but only in one direction. The bedroom

二楼平面图
Second floor plan

1. 露台 Deck
2. 观景门廊 Screen porch
3. 厨房 Kitchen
4. 食品室 Pantry
5. 餐厅 Dining
6. 客厅 Lounge
7. 楼梯 Elevator
8. 化妆室 Powder room
9. 草地阳台 Grass terrace

at the end of the corridor is ensured privacy by its height and an oversized expanse of ceiling to the underside of the deck which extends out from the kitchen. Where the entry level is dedicated to public functions and provided with views and open space, the family oriented rooms on the floor below are provided with a sense of warmth and security.

The clients had expressed at an early stage that they did not care for symmetry. Oshatz removed any need for symmetry by utilizing a radial plan for the house. The plan consists of a series of radiuses, with a number of different centre points. To achieve a logical and harmonious plan throughout the house, each radius is related. The primary radius wraps itself around the contours of the site, following a path that avoids the need to remove any trees. This radius also maintains a convex aspect to the lake, which helps to make the house feel as if it is opening up to the views and the landscape around it. The main axis is centered on a large oak tree that dominates the site. Each subsequent radius responds to the geometry of the others, resulting is spaces that feel at the same time free flowing, and harmonious.

The house was designed to be visually small from the entry way, but to utilize the steep slope of the site to provide maximum views and connections to the lake. The resultant structure emerges from the earth as a series of planes that glide horizontally along face of the hill side. Each plane is unique, and no plane exactly follows any other; they appear to be light weight and are separated only by glass. The roof planes, with a different materiality to the floor planes, spin in their own unique pattern, helping to provide intrigue and complexity to the design. The lightweight and energetic horizontal planes are countered by the vertical stone volumes that appear to grow from the earth and help to anchor the building.

内部
Interior

客厅
Living room

客厅
Living room

女童房
Girls' bedroom

主卧室
Master bedroom

1. 入口 Entry court
2. 客厅 Great room
3. 厨房 Kitchen
4. 卧室 Bedroom
5. 浴室 Bathroom
6. 游戏室 Playroom
7. 洗衣房 Laundry

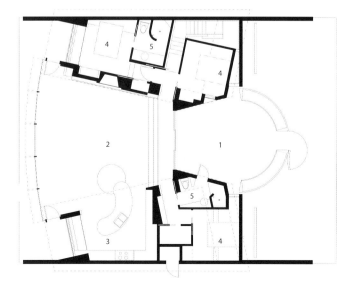

三楼平面图
Third floor plan

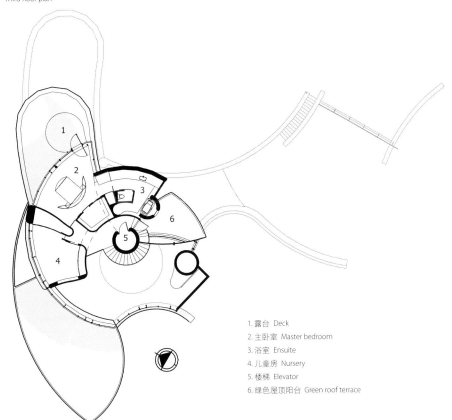

1. 露台 Deck
2. 主卧室 Master bedroom
3. 浴室 Ensuite
4. 儿童房 Nursery
5. 楼梯 Elevator
6. 绿色屋顶阳台 Green roof terrace

Credits

Location: Chenequa, Wisconsin, USA
Architect: Robert H Oshatz
Jnr Architect: Andrew T Boyne
Structural Engineer: Brett King
General Contractor: Signature Builders Inc.
Photographer: Cameron Neilson
Text: Andrew T Boyne

旋风楼梯
Armani Fifth Avenue
阿玛尼第五大道店

| Doriana and Massimiliano Fuksas |

商店空间跨越四个层次,但四层为一个整体,没有明显的界线
The showroom develops on four different levels and it is conceived as a single space, without clear distinctions

一楼和二楼平面图
First and second floor

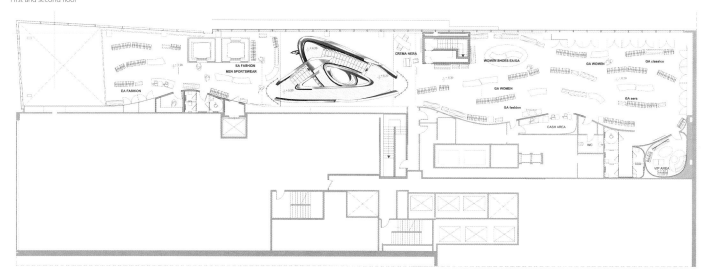

继在香港遮打大厦和东京银座设店之后，Massimiliano and Doriana Fuksas设计事务所规划了阿玛尼在美国纽约第五大道的店面，从而为阿玛尼设店三部曲画上圆满的句号。

阿玛尼第五大道店位于纽约市中心，坐落在第五大道和第56大街交汇处的两座建筑的底部三层。整个展示空间分成四个不同层次，包括地下一层。在空间中，一个如舞动的旋风般的楼梯蜿蜒盘绕其中，将每一层连接起来，形成了令人震撼的视觉效果。

显而易见，这个耀眼的楼梯便是此空间设计的核心元素了，它以辐射状的钢结构为骨骼，塑性材料为皮肤，如巨大的龙卷风从这里"席卷"而过，将每一层的空间连接起来。设计的灵感起源于设计师关于旋风动力的构想，它盘旋于空间中把阿玛尼的世界紧密联系在一起。

楼道如同在风中飘扬的缎带一般，带着人们通往每一层不同的空间，使人们不致迷失在这炫目的几何世界中。

每一层的布局都按照"缎带"的弯曲形态来规划，通过"漩涡"的走向控制着空间的规划。

在这里，无论什么结构都让室内充满了活力，甚至包括外立面的设计；虽然外立面迎合了曼哈顿冷峻锐利的直线条，投射在LED帘幕上的图像和阴影仍旧彰显出大楼的动态感。帘幕除了充当向外的投影仪之外，还表达了对纽约的敬意，无可争议地将现代和动感融为一体。

墙上覆有经过上漆的木板，形成绵延的帘幕，使内部空间充满流动性。

盘旋的楼梯
Rolling staircase

入口
Entrance

二楼和三楼平面图
Second and third floor

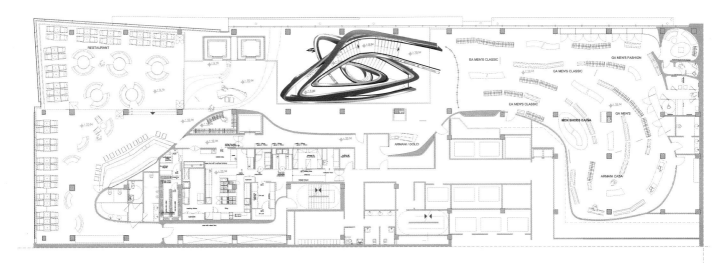

室内进行了区域分隔，试衣间和VIP厅随着楼梯的旋转自然形成。其他的区域，如职员休息室、收银台与Armani Dolci专卖店，都围绕旋转楼梯这个主题元素渐次呈现。这里的照明设计也精心安排，为空间和楼梯的流畅感起到了强调的作用，有着画龙点睛的妙用。

内部设计的每一个要素，从店面到寄存室，从桌子到扶手椅，都跟随着楼梯的运动概念来变化，成为旋风中的一部分。室内的布置处处体现和谐。墙壁、室内陈设焕发的光泽与大理石和楼房顶部的黑色色调形成对照，由此体现出建筑元素之间的互动。

而简易的内部空间与咖啡店/饭店也形成了对比。层折的铜板围合着餐厅，反射并烘托出大楼的色泽和阴影，从而营造出了全新的氛围。在餐厅透过琥珀色的纱帘可以看到第五大道与中央公园一角的壮观景色。

色彩和材料的使用和展厅的其他部分相同，但是意味却不同。这里是娱乐性的空间，地板上的一行灯光引导到餐厅的入口，作为视觉上的幕布，它们让过道生机盎然，很像是戏剧空间。

宽敞的帘幕使街道生机盎然，如同一座剧院拉开帷幕了！

草图
Sketch by Doriana Fuksas

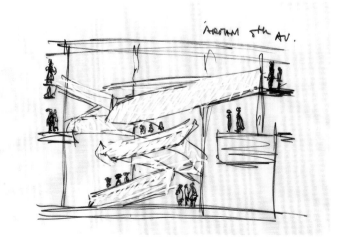

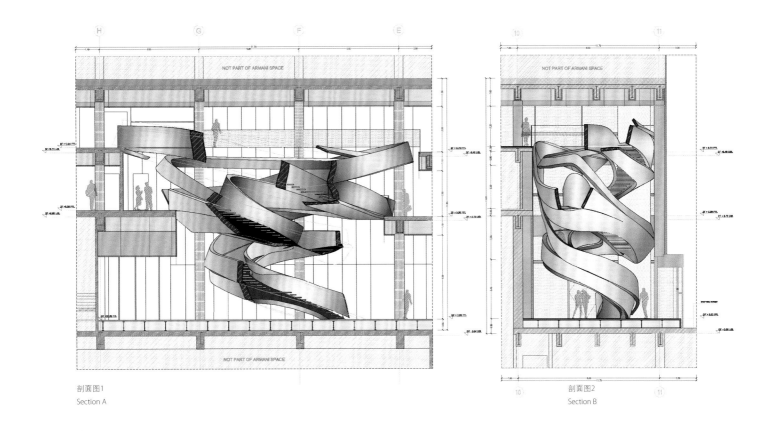

剖面图1
Section A

剖面图2
Section B

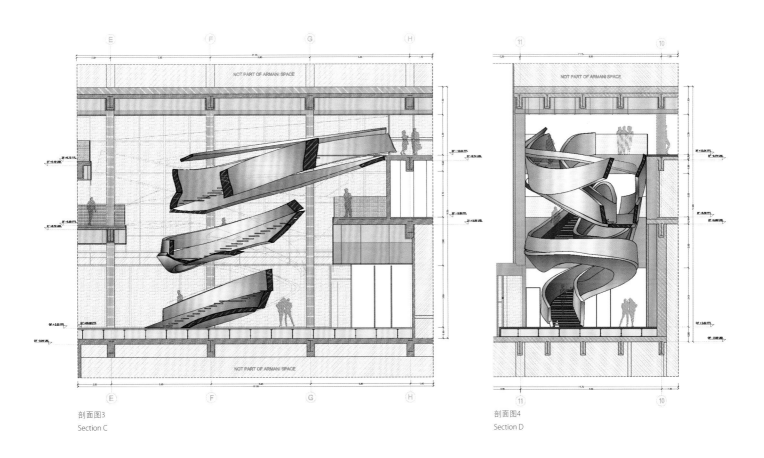

剖面图3
Section C

剖面图4
Section D

楼梯如舞动的旋风般蜿蜒盘绕在空间中，赋予其活力
A space connected with the power generated by the vortex that is the staircase.

After Hong Kong Chater House and Tokyo Ginza Tower, Fifth Avenue
Doriana and Massimiliano Fuksas
Situated in the center of New York, in one of the world's well known streets, the project takes up the first three floors of the two buildings located between 5th Avenue and 56th Street.

Besides the basement, the showroom develops on four different levels and it is conceived as a single space, without clear distinctions, a space in harmony connected with the power generated by the vortex that is the staircase.

The heart of the building is, in fact, epitomized by the staircase, structure in rolled calendar steel (made in Italy) and clad in a plastic layer that highlights its exceptional sculptural presence. It is an entity that is almost impossible to convey in terms of any normal geometric shape that originates from a vortex with great dynamism, surrounded by the different levels that accommodate the Armani world.

The movement of the ribbons that constitutes the staircase, skimming each floor, disenchants the possibility to recognize the geometry.

The general layout of every floor develops according to the different flexures of the ribbons, creating a space controlled by the vortex.

No element is extraneous to the internal dynamism, not even the external façade; even if it is lined up to the rigid orthogonal stitch of Manhattan, simulating the movement through images and shades, projected on a set of LED threads. This screen, besides being the projection to the outside of the internal space, is also a particular tribute to New York City, the inescapable necessity to compare its modernity and its dynamism.

The fluidity of the internal space is rendered by the wall of continuous threads that are realized with lacquered wood panels.

The different rays of bends that outline the threads transform into the spaces and handles for the different product areas. The folding of the threads give hospitality to the dressing-rooms and the VIP hall, also transforming into areas reserved for staff, cash desks, or special product areas such as Armani Dolci. A particular importance is given by the enlightenment that defines, characterizes and emphasizes the bends of the walls and of the spaces, highlighting the different functions of the general layout.

Every element of the internal design, from the shop floors to the

蜿蜒的楼梯引导着来客在阿玛尼的世界里徜徉
The rolling staircase leads visitors in the Armani world

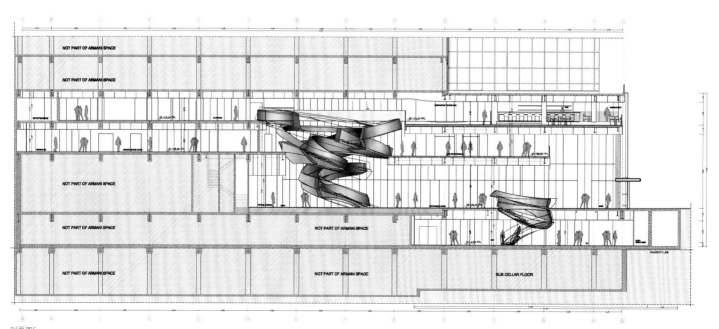

剖面图5
Section E

楼梯以辐射状的钢结构为骨骼,塑性材料为皮肤
The staircase is made from rolled calendar steel and clad in a plastic layer

storage, from the desks to the armchairs, follows and satisfies the movement concept generated by the staircase, becoming a part of the same vortex. The disposition and the route that it implies outline the harmonic layout.

There is a confliction between the shine of the walls and of the furnishings and the noir of the marble and of the ceiling, therefore emphasizing the areas that are there for interaction.

There is also a confliction between the apparent simplicity of the internal space and the cafe/restaurant giving an advanced notice of the elevator's entrance. The refolded bronze that covers them acquires and reflects the colors and the shades giving a glimpse of the new atmosphere. From the restaurant, filtered by an amber veil, there is a splendid view of 5th Avenue and the end of Central Park.

The colors and the materials utilized are the same as the rest of the showroom, but the suggestions are new and different. The space becomes recreational, a line of light on the floor that leads to the entrance of the restaurant, underlining the sensuality of the bends of the wall.

A virtual curtain activates the passage and, just like a theater...the show begins!

Credits

Location: NYC, USA
Year: 2007~2009
Area: 2,800 m^2
Architects: Doriana and Massimiliano Fuksas
Interior Design: Fuksas Design
Contractor: Americon Construction Inc.
Lighting: Speirs& Major Associates
Engineering: Engineer Gilberto Sarti
Client: Gruppo Giorgio Armani
Photographers: Ramon Prat, Studio Fuksas

Live Wire | 电线楼梯 |

| Oyler Wu Collaborative |

这个由Oyler Wu Collaborative 设计事务所完成的楼梯旨在发掘南加州建筑学院展览馆的空间潜能，扩展结构和功能之间的联系。这座楼梯构筑了从展览馆一楼到楼上狭小天桥的垂直通道，保证了它的外观和使用品质的同时，考虑了它的功能、用途和性能。它在空间中建立了一种新的运动形式，打破了展览馆原本的密闭特性，使之有效地融入学校的日常生活中。

为了将多种建筑理念融入单一建筑构件之中，设计师为楼梯寻找了一种合适的"建筑语言"。在传统的垂直系统

这个由Oyler Wu Collaborative 设计事务所完成的楼梯旨在发掘南加州建筑学院展览馆的空间潜能，扩展结构和功能之间的联系

Motivated by the desire to occupy the SCI-Arc gallery in a way that exploits the spatial potential of the existing venue, this Oyler Wu Collaborative installation argues for an expanded relationship between tectonic expression and functional performance

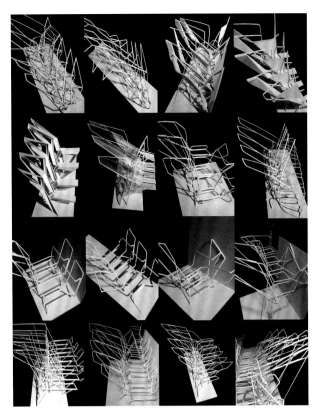

模型
Model

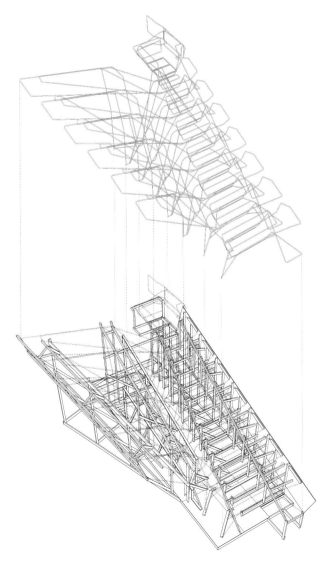

楼梯在空间中建立了一种新的运动形式,打破了展览馆原本的密闭特性,使之有效地融入学校的日常生活中
The stair establishes a new form of movement through the space that challenges the closed nature of the gallery as a hermetic space for objects, effectively integrating it into the daily operations of the school

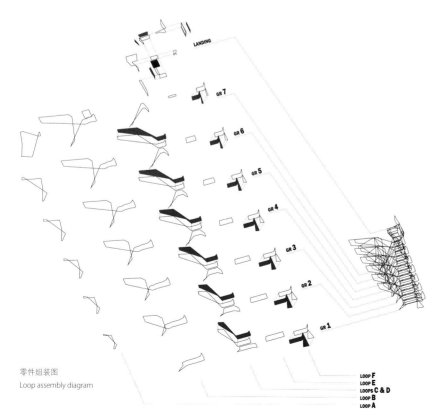

零件组装图
Loop assembly diagram

中,往往会将大量的建筑构件组装在一起,比如,沿楼梯边缘的护栏,一侧墙壁或护栏上的扶手,楼梯表面的踏板,以及用以支撑的纵梁。这些独立的构件通常独立于整个系统,形成建筑外馆。如此相互孤立的建筑模式并不能表达构造的流畅性和建筑的内在含义,必须有对细节更整体化的考虑。这个楼梯使用了约732米的铝金属管和金属棒,由各种类型的复合环组成,将必要的楼梯元素结合在了一起。同时,楼梯表面采用了厚度不一的小型穿孔铝板,打造了一个从台阶到护栏,再到天篷的连续半透明表面。

楼梯作为基本的建筑元素常被认为是纯粹的功能构建,这个装置是对建筑多元概念的尝试,从对光、几何和结构的处理到垂直流通。每一个建筑元素都需要通过不断地改进来达到需要的性能标准。展览馆的长度和台阶的尺寸呈一定的比例,楼梯连着天窗的一侧则与天窗呈另一种比例关系。沿着楼梯向上,几何造型的变化考验了材料的结构承载强度,依靠建造密度来承担负荷。材料的这种密度能提供足够的结构支撑,通过这种形式他们的性能很容易被忽略,取而代之的是给予了空间新的定义。在这种理念下,这个装置似乎扩大了建筑元素的定义,模糊了功能和外形的边界。

细部
Detail

楼梯表面采用了厚度不一的小型穿孔铝板
The stair incorporates faceted perforated aluminum panels of two different thicknesses

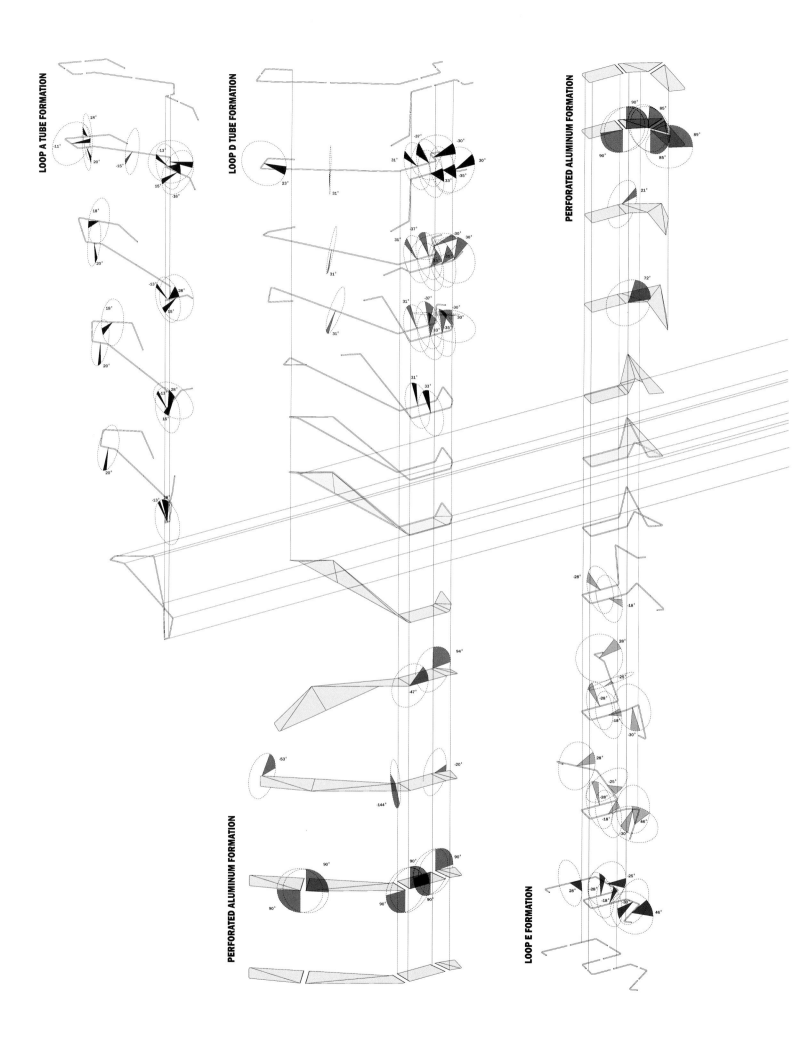

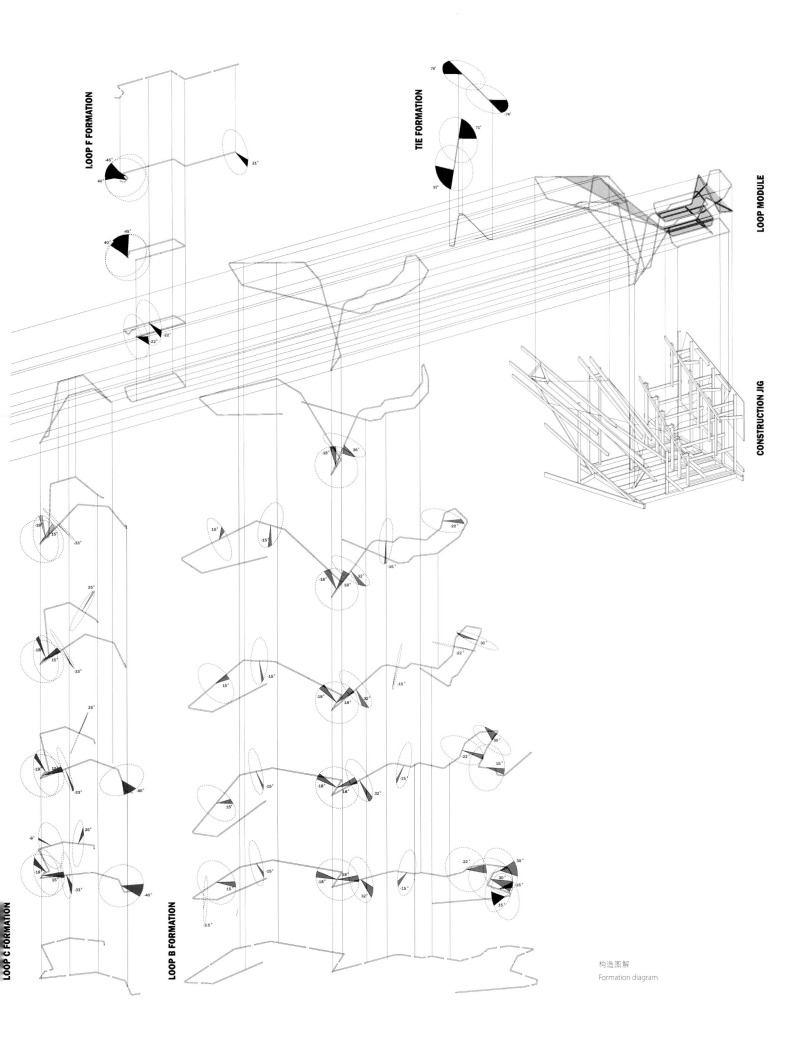

构造图解
Formation diagram

Motivated by the desire to occupy the SCI-Arc gallery in a way that exploits the spatial potential of the existing venue, this Oyler Wu Collaborative installation argues for an expanded relationship between tectonic expression and functional performance. The installation proposes a vertical circulation system linking the floor level of the gallery to the catwalk above. This circulation system, a.k.a. a stair, is equally concerned with its function, use, and performance as it is its visual and experiential qualities. The stair establishes a new form of movement through the space that challenges the closed nature of the gallery as a hermetic space for objects, effectively integrating it into the daily operations of the school. With the intention of bridging multiple architectural ideas within a single architectural element, the stair exploits a tectonic language appropriate to that objective. In conventional systems of vertical circulation, numerous components are assembled together, with each performing a specific function, for example, guardrails provided along the perimeter, handrail attached to adjacent walls or guardrails, tread and risers for stair surfaces, and a stringer for structural support. Furthermore, these individual components often act independently of systems meant to shape architectural experience. This segregated tectonic formula leaves little room for consideration of the kind of fluid spatial and tectonic implications that might result from a more collective consideration of the parts. Constructed of approximately 2,400 linear feet of aluminum tubing and rods, the stair employs a combination of complex loops that perform a variety of tasks as they merge together to form the necessary stair elements. Similarly, the stair incorporates faceted perforated aluminum panels of two different thicknesses to create a continuous, semi-transparent surface from stair tread to guardrail to

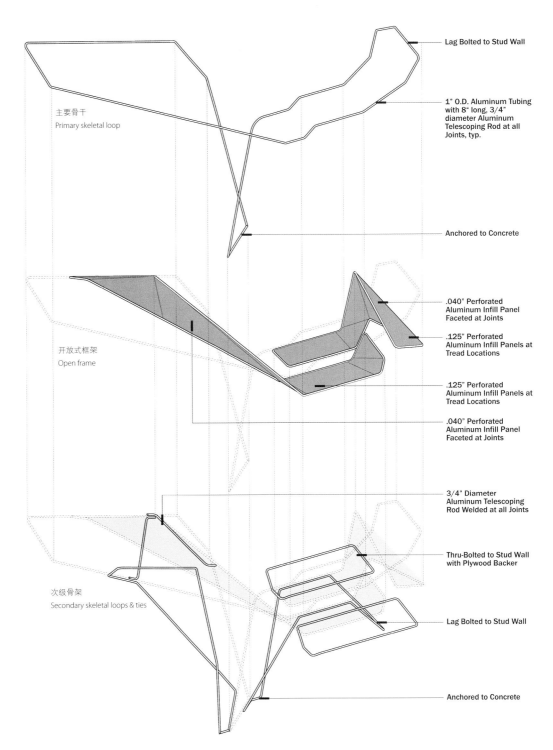

主要骨干 Primary skeletal loop	Lag Bolted to Stud Wall
	1" O.D. Aluminum Tubing with 8" long, 3/4" diameter Aluminum Telescoping Rod at all Joints, typ.
	Anchored to Concrete
开放式框架 Open frame	.040" Perforated Aluminum Infill Panel Faceted at Joints
	.125" Perforated Aluminum Infill Panels at Tread Locations
	.125" Perforated Aluminum Infill Panels at Tread Locations
	.040" Perforated Aluminum Infill Panel Faceted at Joints
次级骨架 Secondary skeletal loops & ties	3/4" Diameter Aluminum Telescoping Rod Welded at all Joints
	Thru-Bolted to Stud Wall with Plywood Backer
	Lag Bolted to Stud Wall
	Anchored to Concrete

这个楼梯使用了约732米的铝金属管和金属棒,由各种类型的复合环组成,将必要的楼梯元素结合在了一起
Constructed of approximately 2,400 linear feet of aluminum tubing and rods, the stair employs a combination of complex loops that perform a variety of tasks as they merge together to form the necessary stair elements

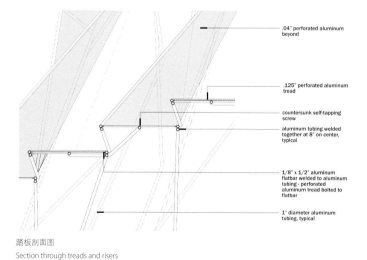

踏板剖面图
Section through treads and risers

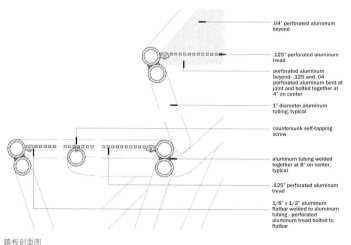

踏板剖面图
Section through treads and risers

canopy.

Often relegated to pure functional use, the fundamental architectural element presented in this installation is a testing ground for weaving together a multitude of architectural ideas, ranging from the manipulation of light, geometry, and structure to, of course, vertical circulation. Conceived of first as a series of light modulators, each architectural element requires a progressive manipulation in order to negotiate the required performance criteria. With the length of the gallery and the size of the treads providing a scale to one side of the intervention, the opposite side reaches up toward the clerestory windows at a dramatically different scale. As the stair moves upward, the geometry takes on a transformative quality that pushes the structural limits of the material, relying on the built-up density to carry the load. As much as this density of material is meant to provide structural support, it is recognized that it is within these areas that their performance is most easily forgotten, giving way to the spaces they define. It is at this conceptual intersection that the installation is intended to provide a more expanded definition of architectural elements, one that knows no boundaries between the simple functions they perform, and the more intangible results that they evoke.

展览馆的长度和台阶的尺寸呈一定的比例,楼梯连着天窗的一侧则与天窗呈另一种比例关系
With the length of the gallery and the size of the treads providing a scale to one side of the intervention, the opposite side reaches up toward the clerestory windows at a dramatically different scale

Credits

Principal Architects: Dwayne Oyler, Jenny Wu
Project Leaders: Ming Jian Huang, Matt Evans
Design Team: Fayez Ahdab, Phillip Cameron, Ming Jian Huang, Huy Le, Erik Mathiesen, Dwayne Oyler , Jenny Wu
Engineering: Matthew Melnyk, BURO HAPPOLD

西立面
West façade

Split View
|UArchitects|

形散而神聚——错位住宅

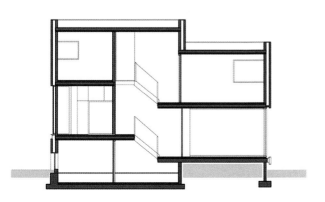

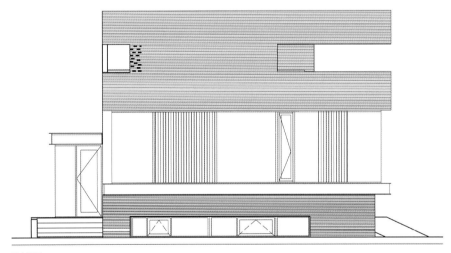

西立面图
West elevation

项目介绍

由于居住状况发生变化，这户家庭意欲修建一座新住宅，并在新住宅中能体现这种变化。一方面两个孩子和家长都希望拥有自己独立的生活和居住空间，另一方面他们又不想分离开来，互相还要有必要的交流和联系。同时住户还要能欣赏周围的风景。从某种意义上说，这个新建的住宅是在诠释一种正在经历变革的家庭关系，这样的变革也会在许多其他的荷兰家庭发生。

客户希望屋内拥有充足的采光，在不同的季节带来不同的体验，而他们希望这种体验将体现在新住宅的每一个角落里。

住宅外部展示出一种可见的错层排列，从室内外的不同角落可以看见住宅的不同功能区，这里的视野拆分也是精神拆分，因为一个居住在传统住宅里的人只会在传统的封闭的空间中才会有稳定的精神和记忆状态（引自约书亚·富尔的书籍）。

这个主题曾经出现在约书亚·富尔（一位美国自由记者）的书中，他提出如果一个人离开他阅读或是学习的封闭空间，他刚刚的记忆会很快被丢弃，但是当他再回到同样的空间中，这些记忆又会迅速浮现。所以可见我们的思想和记忆是按照不同的房间或空间来分类的。

该房间将是一个有趣的思想和行为实验，因为它包含一系列不同空间的交叉和连接。我们时常扪心自问，这样错层的住宅的内部布局拆分会给人们的思维、记忆带来何种影响，会否引起断裂错开的记忆和思维，要知道这

种现象在现代社会的记忆中十分多样，且依存于如今这多变和动态的社会之中。记忆和行为会不会被这种住宅内部的穿透性和开放联系所影响呢？这套住宅的室内和室内的功能，从室外都能进行解读，同时住宅的每一侧都呈现出自己独特的性格，有时封闭，有时敞开，有时友好，有时坚决，有时似乎没有任何反应，但有时候它又大开，暴露自己的内心。

材料

住宅位于两条马路交叉口，楼梯主要采用黄杨木建造，连接住宅各层。设计师计划将该住宅作为周边的地标，用钢梁为不同的楼层标注记号，同时在荷兰，这些钢梁也可以作为路边护栏。其他材料有双色砖（长×宽×高=240×90×40毫米），水平摆放的水泥砖和不同规格的垂直木料（Plato木头：经过精细加工的木料），以及镀锌的钢梁，用于标记住宅错位的初始位置。而砖石则被漆上了稀松平常的黏土。设计师意图在外立面设计中寻求平衡，表述错位理念的主题。

南轴测图
South isometric

从一楼向上望去
Seen from the first floor

起居室内安装了落地门窗，通透明亮
Natural light shines through floor-to-ceiling doors and French windows into the living room

楼梯细部
Staircase details

Project

This family which wanted to build this house was motivated by the change of their living situation and wanted to express this also in their new house. The two children of this family wanted to have their own living surrounding but also wanted to be in close contact with the parents. On the other hand the parents wanted to have more private zones in the new house but also be in close contact with their children. Every floor of the house had to have an overview of the surrounding. In a certain way the new house is a translation of the family situation which is undergoing a transformation like many other families in the Netherlands.

The family wanted a house which is flooded with light, which is experienced each season in a different way and they wanted to experience the surrounding from different locations in the house.

From the outside this house gives a readable split level. A glimpse of the various functions of the house can be seen at different corners from outside and inside the house. For the term "split view" we could also use the term split mind, because the memory of the users of an average traditional house is at its best in traditional closed spaces (book of Joshua Foer).

This theme mentioned in the book of Joshua Foer is as followed: If the person leaves the space where he has learned or read, the thought is easily forgotten by leaving the closed space and when the person returns to the same space, then this person remembers again the thought in question. Our mind/memory is categorized in rooms/spaces.

This house will be an interesting thought and behavior experiment because here are various crossover and open connections between different spaces. We question ourselves which effect this split view house will have on the mind and the remembering of facts and thoughts of the users of this house and will it result in a split mind/memory which is a modern variant is of our modern time of remembering and acting in this rapidly changing and liquid society. Will the task of remembering and acting be influenced by the see- through and open connections in this split view house?

The interior of this house and the use is readable at the outside of the house and each side of the house is reacting in his own peculiar way, one time closed and the other time open, sometimes friendly and the other time hard-minded,

sometimes unanswered and the other time it lights up and shows the inside.

Materialization

The staircase is made of wood (Yellow Poplar) and connects the several levels of the house. The house stands at the cross point of two roads. We proposed to lift the house up from their surrounding and to mark the different floors levels with steel beams which are also a reference to the road protection beams at the side to the roads in the Netherlands. The other materials are the two colors bricks (lxbxh= 240x90x40 mm) strong horizontal deep lying cement and vertical wood elements with different size (Plato wood: special modified wood), and the steel beams are galvanized and marks the beginning of the split in the house. The bricks are specially made in the color with certain amount clay. We try to find a balance in the facades and we try to express the split view theme at the outside.

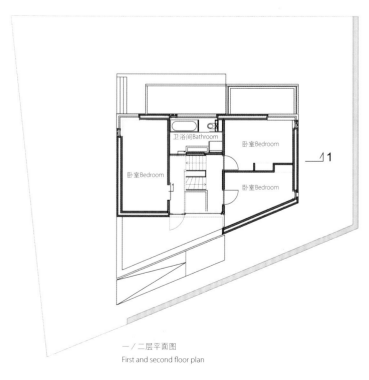

一／二层平面图
First and second floor plan

Credits

Location: Voorste Akkerweg 2, Mierlo-Hout, the Netherlands
Year: February 2012
Gross Floor Area: 250 m²
Contractor: Jansen&Vis
Architects: UArchitects
Founders and Owners of UArchitects: Misak Terzibasiyan & Emile van Vugt
Advisor Construction: Vervest Constructief
Advisor HVAC: Volantis
Architecture Photographer: Daan Dijkmeijer
Client: Family van Renterghem

设计师对光线与色彩的巧妙利用,让室内每个角落都光亮而洁净
The quiet interior space suddenly cheers up with rich natural light

| Natalie Dionne Architecte |

Maison E3
宁静以致远——E3住宅

这座E3住宅的内部空间如同不沾尘世的修道院那样宁静。

多层式E3住宅，其设计理念强调了光与空间的奢华，使其即刻变为一个活跃、有创意的家庭聚会场所。

"E3"住宅是由Natalie Dionne Architecte建筑事务所命名，该团队由Natalie Dionne 与Martin Laneuville 夫妇二人组成，共同完成了E3横截面的规划。他们的方案深受来自维也纳的现代主义设计师阿道夫•路斯的"体积规划"理论影响，这一点可以在错位的房间布局和楼层间自然的过渡区中察觉一二。

E3住宅地处繁华之境，那里靠近蒙特利尔熙熙攘攘的让•泰隆市场。该住宅专为那些深谙这喧闹区域的家庭而设计——这些年轻的父母大多

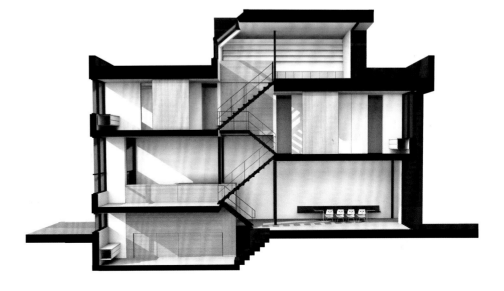

在剧院、影院以及电视台工作。

整个住宅外部如同一个简单的平行六面体,东、西方向的墙壁带有窗户,南、北方向则是共用墙。设计师充分利用地势的走向,使这栋多层的住宅内部光线充足,宽敞明亮。住宅室内有12米高的中庭,将房间分为前后两部分。内部共有六间房间,分布于中庭两侧,各层之间由一个顶部带天窗的楼梯相连。

住宅两侧设有大窗,而太阳光则透过位于中心处的天窗照射到屋内深处,形成光影交错的美景。住宅内部环境随着日夜和季节的更替而变化。每间房内窗户外面都安装有木制百叶帘,早晚的阳光透过百叶窗照射进室内,形成光怪陆离的景象。为了确保自然光照,窗户呈对称式设计,且都呈打开状,而天窗也是如此。同时,百叶窗可以遮挡夏日酷热的阳光。

住宅名曰"E3",因为它是沿着大楼的横截面而规划的,大楼的交互结构犹如字母"E"以及颠倒过来的字母"E"一般。每一间起居室和房间都位于不同楼层。住宅内的六间房都通过楼梯相通,最终通往夹层处的工作室和一处可以观赏皇家山的观景平台。屋顶将种植一小片薰衣草,为房间带来生气。

卧室朝向住宅中央敞开,中间经过了宽敞的旋转门或滑动门。一扇扇玻璃门不仅在视觉上增大了空间,同时带来无限空间的深邃感。关上门,空间自然而然地消失,房间又重回私密、安静的态势。

住宅主体采用统一化的建筑元素和表皮,打造出栩栩如生、雕塑般的空间,明亮通透的中央楼梯与别具一格的厨房都用钢铁和胡桃木打造,而橱柜、壁橱和其他储存设施全由取自于枫树的木制胶合板构建,垂直分布,形成了一处多功能、多层次的大型构件。前后外立面与屋顶采用咖啡色耐蚀胶合板,一侧壁龛式内凹的设计勾勒出建筑优美的姿态。住宅整体交替呈现不同材料简单的色调。经打磨的水泥板、天然钢材、原木以及蓝色瓷砖,这些材料过渡自然,共同造就了这座如画般的建筑。

厨房内部装饰采用钢铁和胡桃木完成,落地式玻璃窗连通开放式阳台,为内部引入自然光线
The kitchen is clean cut with steel and walnut finishes, with floor-to-ceiling windows leading to outside deck

楼梯拐角处放置了舒适的沙发,可享受窗边美丽的阳光
The stair landing by the window offer a comfort corner for enjoying beautiful sunshine with sofas

The E3 House offers an interior environment that is almost monastic in its uncluttered serenity. The delineation of the volume through the multi-level concept accentuates the luxuriousness of light and space. This urban house is at once a living, creative, and gathering place designed for an inspiring family. The E3 House gets its name because Natalie Dionne Architecte, the husband-and-wife team of Natalie Dionne and Martin Laneuville, organized the design in cross-section, using a strategy that recalls modernist Viennese architect Adolf Loos's influential Raumplan, with its staggered rooms and fluid transitions between floors.

The E3 House is located in a bustling neighborhood near the very popular Jean-Talon Market in Montreal. It was designed for a family with a deep attachment to the neighborhood; the parents work in theatre, film, and television and the

children are young adults.

The exterior geometry of the residence is a simple parallelepiped defined by two largely fenestrated walls on the east and west sides and party walls to the north and south. The orientation of the lot inspired the design of a multi-level house that enables natural light to penetrate. The floors are staggered on either side of the 12-metre-high central atrium that divides the house into two volumes, front and back; a staircase, topped with a skylight, links the different levels.

The large windows situated at both ends of the house and the central skylight allow the sun to reach deep into the interior to create ever-changing plays of natural light and shadow. Thus, the interior environment modulates according to the time of day and the season. Large wooden shutters slide in front of the windows in each room to filter the light at dawn and sunset. To ensure natural

ventilation, the windows are facing each other and they all open, as do those in the skylight. The shutters also protect the house from summer heat.

The house is called E3 because it is organized around the cross section of the building. The cross-sectional drawing resembles a capital E linked to a backward E by the intersecting staircase. Each living space and room is on a separate level. The staircase links the six rooms in the house, leading ultimately to a mezzanine studio and a terrace with a view of iconic Mount Royal. On the other side, the skylight looks out on a green roof system, where a small field of lavender will be planted.

The bedrooms open onto the central space through wide pivoting or sliding doors, which expand the space, allow light to enter, and create a depth of perspectives. When the doors are closed, the space is withdrawn, allowing for isolation, privacy, and contemplation.

The volume is structured by integrated architectural elements and finishes that contribute a graphical and sculptural quality to the space. The central staircase, light and airy, and the impressive kitchen island both feature steel and walnut. Cabinets, wardrobes and storage spaces, made with maple-veneer plywood, are vertically arranged to create multi-functional, multi-level monoliths. Outside, marine-grade plywood stained to a dark espresso color lines the walls and ceilings of large alcoves to mark both front and back entrances. The project's program includes a code of materials with simple, repetitive colors: polished concrete, natural steel, wood and blue tiles. The interplay of these materials creates stunning graphical compositions that resemble abstract paintings.

楼梯口顶部天窗为内部带来无限光亮
The skylight on the top of the stairways brings light to the interior

钢铁扶手
Steel handrails

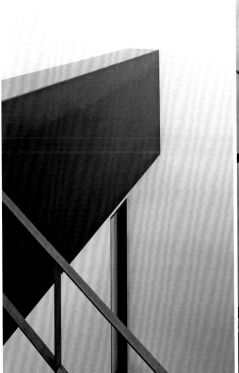
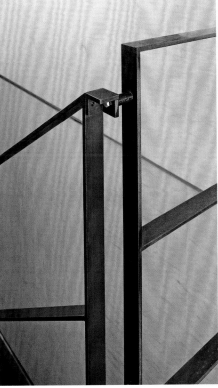

Credits

Location: Montréal, Canada
Completion Year: 2012
Area: 300 m²
Architects: Natalie Dionne Architecte
Project Manager: Natalie Dionne
Design Team: Natalie Dionne and Martin Laneuville
Contributor: Claude Lafrance
Engineer: Alain Mousseau of Calculatec
Contractor: Pierre Aubin
Photographer: Marc Cramer

Stewart Museum at St. Helen's Island
圣海伦岛斯图尔特博物馆

| Les Architectes FABG |

透明玻璃将建筑与周边环境融为一体
The museum merges with surroundings via transparent glass

圣海伦岛兵工厂(斯图尔特博物馆)建于1820至1824年间，以前是安放弹药和枪支的仓库，如今被改建成了一座军事博物馆。斯图尔特博物馆内展示了30 000余种"新法兰西"时期的物件和艺术品。建筑师对场地新规划仔细斟酌，并对大楼功能与技术可行性进行了深入研究。

博物馆翻修项目是基于以下标准进行的：

1. 统一的连体建筑结构；
2. 促进楼内循环设计，提升旅客的观感；
3. 改善气候舒适感；
4. 提高行人和行李的安全性：改善警报系统、灭火器以及紧急出口；
5. 为残障人士进出提供便捷；
6. 机械和电力系统在高能效、结构一致以及安全方面的改进。

斯图尔特博物馆的庭院处设计了一部电梯以及一座建筑楼梯，从而方便外界人士出入。楼梯往返穿梭于三个楼层之间。

电梯四周被镜面围合，营造出万花筒般的建筑效果，将新建筑和周边浓厚的历史氛围紧密地结合起来。

玻璃与钢铁描绘出楼梯干练的线条
The clean-cut outline is created by glass rails and steel armrests

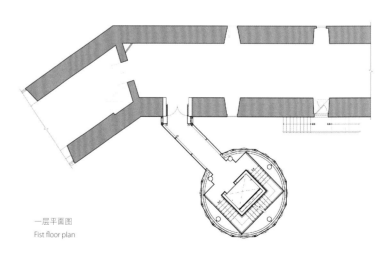

一层平面图
Fist floor plan

L'Arsenal du Fort de l'Île Sainte-Hélène, a former warehouse for ammunition and guns, was built between 1820 and 1824 and was converted into a military museum in 1956. The Stewart Museum has a large collection of over 30,000 objects and artifacts of Nouvelle France and the European influence in North America.

For this project, the architects revised the master plan of the site development and did feasibility study which included a functional and technical program.

The museum's transformation program was developed to meet the following requirements:
1. Grouping functions which were similar;
2. Facilitate the circulation and improve the visitor's experience;
3. Improve climatic comfort;
4. Improve the safety of people and goods: alarm system, sprinkler, and emergency exit;
5. Improve accessibility for disabled people;
6. Improving the mechanical and electrical systems: efficiency, conformity and security.

The project involves the installation of an elevator in the courtyard and an architectural staircase to improve universal accessibility and the circulation through the museum's three levels.

The elevator is covered with mirrors to create the effect of a kaleidoscope which binds the new construction to its historical context.

这座透明的楼梯成为庭院一道亮丽的风景
The transparent staircase has been part of the yard

连通楼梯与博物馆之间的走廊
The passage connecting the staircase and the museum

Credits

Location: Montreal, Canada
Year: May 2011
Area: 2,973 m²
Architects: Éric Gauthier, André Lavoie
Team: Marc-Antoine Fredette, Dominique Potvin, François Verville
Photographer: Steve Montpetit

内部楼梯螺旋而上,直至屋顶,增强了垂直方向的空间感
The sense of vertical flow between levels is promoted by the spiraling half-oval steel stair terminating in a small skylight

Greenwich Village Townhouse
"钢"柔并济
格林尼治村联排式别墅区

| Gates Merkulova Architects, LLP |

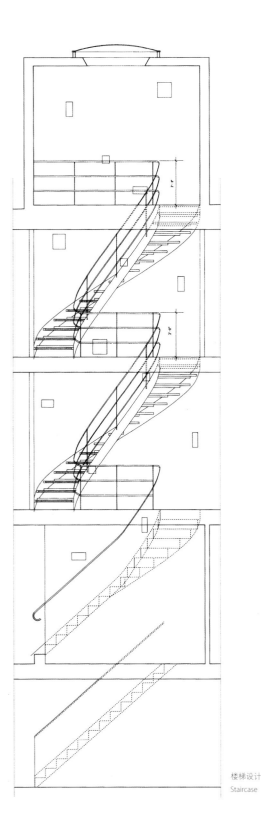

楼梯设计
Staircase

椭圆状螺旋的楼梯
Half-oval steel stair

这片联排式别墅区位于格林尼治村，建于19世纪早期。它的结构在纽约这座城市显得十分怪异：显然，临街方向已有一个入口，但主入口却穿过一个私家后院，朝向四间联排式别墅的公共庭院，与街道不通（中间隔着一道装饰性的栏架）。

由于周边环境恶劣，住宅区需进行重建，新的楼房被嵌入式地建在三堵原有砌石墙之间。加上住宅区入口不止一个，这样的结构更显街道和花园之间区域的二重性。位于底楼租赁单元房入口处的街道外立面在保留其历史风貌的基础上进行了修缮。而花园古旧的砌石墙外立面被铝合金与玻璃外皮所取代，这样的外立面仅仅在私人庭院处露了峥嵘，极具现代感。大楼庭院四周的墙壁上爬满了葡萄藤，一道布置了盆栽的楼梯矗立在墙壁一

侧，通往建筑入口。

房间面积较小，而传统的联排式别墅的平面规划往往会导致室内空间阴暗狭窄，与现代生活方式格格不入。而格林尼治村的联排式别墅区内，每一个楼层都以复式的形式建造，显得空间既开放又灵活多变。可移动式半透明玻璃用于分隔空间。玻璃后门墙处竖立着的巨大的滑动门和位于二楼的阳台使得内外空间十分通透。钢质楼梯呈螺旋状上升，直至屋顶，增强了垂直方向的空间感，屋顶半椭圆形的天窗则为楼梯口引入光亮，更显得室内宽敞明亮。

屋顶平台被规划成一座私人花园，主要被划分为两个松散的区域：位于庭院末端独立的室外餐饮区，以及一处位于街道一侧的更开放的休憩区。而诸如铝合金镀层的楼道隔板和砌石墙包裹的烟囱之类的大型建筑构件勾勒出屋顶的外观。

业主是位摄影爱好者，她极力将自己的作品与住宅的设计结合。照片被夹在薄薄的玻璃之间，显得清晰而透明，这些照片被固定在实体墙与隔断屏风上；这些墙壁围合着楼道，形成了一处垂直的私人作品展示画廊。

直到这栋住宅拥有了新的业主，才进行了第二次重建。二次重建规划的目的是将屋子打造成独栋别墅。住宅区的修建包括了花园外墙、内部的楼梯及背光照片的艺术画廊。住宅区平面布局、通往内部的走道以及内部的饰物被完好地保留着，并根据新业主的意愿进行了适当的修饰。花园的楼道和入口处依旧保留，而花园本身则被重新设计，成为了一处更为正式的庭院。

屋顶小天窗
Small skylights at the rooftop

钢筋楼梯之钢筋美
The figure of steel staircase

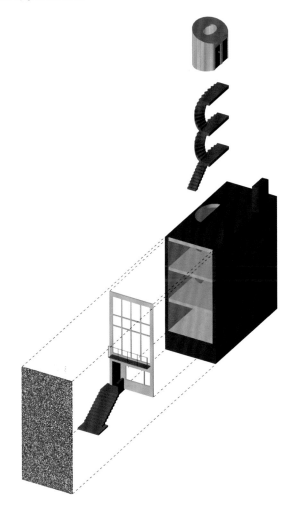

View of the screen partition

Spiraling space

This townhouse in Greenwich Village was built in the early 19th century. Its setting is quite unusual for New York: although there is a street entrance, the main entry is through a private rear yard opening to a court shared by four townhouses and separated from the street by an ornamental gate.

Due to its deteriorated condition, the house had to be virtually completely rebuilt, with new structure inserted within three of the existing masonry walls. This, combined with the unusual entry sequence, has allowed for exploration of the duality between the street and the garden. The street façade, where the entrance to the separate lower-floor rental unit is located, has been restored to preserve its historic character. The old masonry wall of the garden façade has been replaced with a taught aluminum and glass skin which is revealed, as a modern surprise, only from within the private courtyard. The courtyard is bounded by vine-covered walls of the adjacent buildings and is dominated by a generous, landscaped entry stair.

The house is small, and the traditional townhouse floor plan would have resulted in cramped, dark and rigidly defined spaces unsuitable to the modern lifestyle. Instead, each floor has been treated as a "loft", creating openness and flexibility. Translucent materials and sliding partitions have been used wherever separation between spaces is required, while the glass rear wall, with its large sliding doors and a second floor balcony, has allowed the outdoor space and light to flow into the interior. The sense of vertical flow between levels is promoted by the spiraling movement of the half-oval open-riser steel stair terminating in a small skylit room at the roof level.

The roof terrace has been developed as a private garden, loosely divided into two areas: a secluded outdoor dining area at the courtyard end, and a more open sitting area on the street side. The space is defined by the over-scaled architectural elements of the aluminum-clad stair bulkhead and the masonry fireplace chimneys.

被盆栽与葡萄藤围绕的外部入口楼梯
The exterior living staircase

花园一侧房间
The room by the garden

The owner of the house was a photographer, and she collaborated extensively in incorporating her work into the design of the house. Transparencies of her images, laminated between sheets of glass, were used as permanent screen partitions and as removable rear-lit glass panels inserted into the wall enveloping the stair, creating a vertical gallery for her work.

A second renovation of the townhouse was undertaken in 2006, when it was sold to a new owner, primarily with the purpose of converting the house to a single-family residence. The architecture of the house, including the garden façade and the interior stair with its gallery of back-lit images, was kept in place. The house plan and the approach to the layout and treatment of the interior spaces have been retained as well, with some modifications designed to meet the new owner's needs. The garden stair and the entry sequence also remain, while the garden itself has been re-designed as a more formal court.

Credits

Location: Greenwich Village, New York, USA
Site Area: 92 m²
Construction Area: 335 m²
Designer: Gates Merkulova Architects, LLP
Photographer: J.B. Grant (GVT.01,.05,.15.16), Gates Merkulova Architects, LLP (All Others)

东视角
View from the east

树屋建在一个树木茂盛的基地上，周围是有着百年树龄的树木和一条季节性溪流。

由于存在洪水隐患，可建造的面积非常小。另外，此地是一个百年冲积平原，为保护基地上所有树木，住宅被设计呈垂直状。

垂直走向使"树屋"成为可能，在上面两层可以看到不可思议的天空景象。客厅在二楼，主卧和浴室占据了整个三楼。这些凌驾于树枝的空间让人感觉自己似乎成为了天空的一部分。其余的空间包括车库旁的备用卧室和二楼的书房。

内部的楼梯由Sander建筑事务所设计，在客户工作的地方焊接完成。它们由1.3厘米厚的铝板制成，每个单元两个踏板。这部楼梯为客厅引入雕塑元素，与双层楼高的暗熏衣草色石板壁炉墙相合。

外面两部楼梯连通室内。住宅前方入口处有三棵树，一部悬臂式楼梯通往室内。这部楼梯有篦板台阶，可以看到下面的溪流，同时阻止积雪。东立面从底楼通向屋顶的螺旋楼梯创造了一个在屋顶嬉戏的室外空间。

从街道向南方望去
View from the street to the south

Tree House ｜树屋｜

｜Sander Architects｜

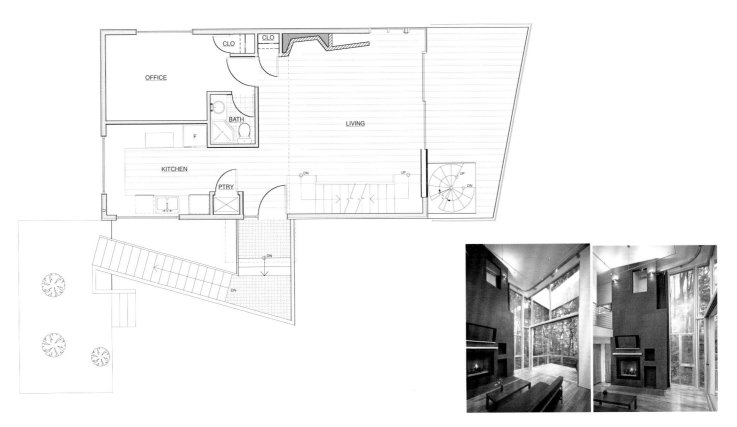

楼梯正面,可以看到镀钢踏板
View of the front stairs showing galvanized steel bar grate treads

Sander事务所设计并由客户焊接的定制铝台阶
Custom aluminum stairs designed by Sander Architects and fabricated by the client

Tree House sits on a wooded site surrounded by century-old trees and a seasonal stream.

As a result of the potential for flooding, the buildable area is quite small. The combination of the constraints of the hundred-year flood plain on the building pad, with the decision to preserve all the trees on the site, decided the vertical form of the house.

This vertical form allows the house to sit in the trees, and to have marvelous views of the canopy from the two upper floors.

Horizontal windows encircle the house and provide select views into the landscape. In contrast to these small views, a great wrapping window in the double-height living room provides a dominant diagonal focus for the house, and provides dramatic views into the deep woods.

The main living area is on the second level, while the main bedroom and ensuite bathroom occupy the entire third level. These raised spaces give one the feeling of being part of the canopy itself. Other spaces include a spare bedroom at the garage level and a study on the second floor.

The inside stairs were designed by Sander Architects and fabricated at the performance stage shop where the client works. They are made of 1/2" aluminum plate, with two treads in each unit. They create a sculptural element for the dramatic living room and provide a counterpoint to the double-height fireplace wall that is clad with dark lavender slate.

Two exterior stairs access the house. A front entrance stair, cantilevered away from the façade, enters through a trio of trees. This stair has bar grate steps that provide a view of the stream below and also discourage snow build-up.

On the east façade, a spiral stair wraps up from the ground level to the roof deck which doubles as an outdoor living space for tree top entertaining.

Credits

Location: Wilmington, Delaware, USA
Floor Area: 232 m²
Site Area: 30 ha
Photographer: Sharon Risedorph Photography

东北视角
View from the northeast

New Sincretica Offices

| Giovanni Vaccarini Architects |

全新一体化办公室

该项目是对一幢建于19世纪的居民楼进行翻新，居民楼位于历史悠久的朱利亚诺瓦旧城区，如今大楼将被改造成办公空间。

设计师将办公空间视为外界环境的自然延续，完全不同于历史建筑的特色，这便是该办公室的规划理念。

为了达到这一目的，设计师首先将入口的高度提升了一倍，如此一来，光照范围增大，在办公室的一楼和二楼之间建立了延续感。

两部楼梯为办公室内部添色不少。楼梯由金属架支撑，木板台阶看上去好像是悬浮在空中，下置"穿孔金属板"加以支撑。楼梯的栏杆由一根金属管构成，管道构造了一处由一楼通往阁楼的宏伟"线条"。

项目规划的另一个重点建筑构件是门槛。

入口大门采用的建筑元素是金属板和磨砂玻璃，临街而开，在入口处有座小亭，四周被木板围合，一侧被漆成了纯白色，内部设电话和邮箱。门槛外部的石灰华地板也覆盖了整个一楼，使内外保持了延续性。

金属框架支撑着楼梯踏板
A metal frame supports the treads

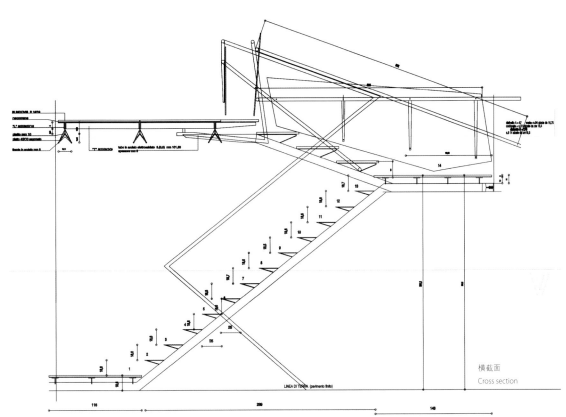

横截面
Cross section

穿孔的金属网支撑着木质楼梯踏板
The wood treads are supported by a perforated metal cabinet

The project involves renovation of a residential building in the nineteenth century belonging to the historic fabric Giulianova Alta (TE) to be used as an office.

The idea is to think of space as a smooth succession of environments strongly separated from the architectural nature of the historic building.

To achieve the architectural idea of continuity is primarily the double height entry space that allows a perception continues between the ground floor and first floor thanks to the lightness of the scale.

The interior space is characterized by the double flight of stairs. It consists of a metal frame on which the treads are almost suspended wood, supported by a perforated metal cabinet. The handrail is a metal tube that draws a strong line from the ground up to the loft where it becomes parapet.

The other architectural element on which the project focuses is the threshold. The entrance door metal and opaline glass set back from the street front, creating a niche drawn white on one side by a wooden panel that includes the telephone and the mailbox. The travertine floor outside the threshold extends over the entire surface of the ground floor, creating a material continuity between inside and outside.

木质踏板
Detail of wood treads

金属管构成的栏杆
The handrail formed by a metal tube

Head Office for Piper Heidsieck and Charles Heidsieck Champagnes

| JFA / jacques ferrier architectures |

拍谱海锡克&查理海锡克香槟总部大楼

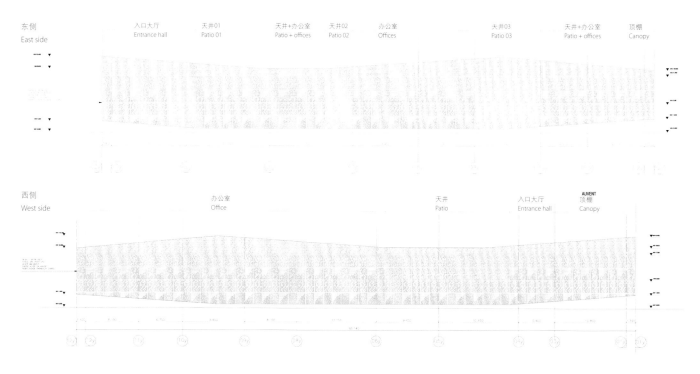

幕墙夜景
Panel wall by night

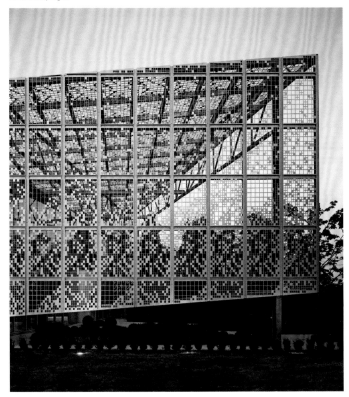

拍谱海锡克&查理海锡克香槟酒庄的新办公大楼包括建造新的生产区，扩建一个新的葡萄种植产地和一个展览区域。建筑中所有的元素通过同一种风格呈现，自然地融为一体，同时使得建筑既现代又不是很突兀，还能引起人们对往昔的回忆。建筑整体从整个景观中凸显出来，形成了类似灯塔的复合结构，改变了整个种植园的面貌。

酒庄建筑结构必须最小化，因此该建筑精致的金属结构需要简洁、高效，并且在这之中体现典雅和高贵。

新的办公建筑被设计成一个包在网格内的玻璃结构，这样在向天空延伸的时候，给人的感觉是慢慢在非物质化——它可以给人们一种类似香槟泡沫不断消失的印象。白天，网格同发散出的金色光芒与不断变化的天空之色融合在一起；夜间，建筑内部光线照射出来，使得整个建筑成为光源。

亲水硬地散步道上方的大型顶棚延伸至被景观包围的停车场处，似乎在迎接着来客。带有顶棚的散步道连通至大型接待厅，这里，从屋顶射下来的光经过金属框架和金属板过滤后，在大厅弥漫。透过彩色玻璃窗的光线为室内增加了更多的色彩。

建筑的中心是透明的，可供欣赏外部景观，同时也形成了灵活的内部空间布局。内部通过三个天井花园划分成了四个馆，自然元素的加入让整个空间环境显得更加宁静。

该建筑融合了香槟酒庄建筑的形式和人类活动对舒适度的要求。同时，它还强调用途和与周边环境的关系。可以说，它既满足了功能需求，又尽显典雅大方。

纵剖面
Longitudinal section

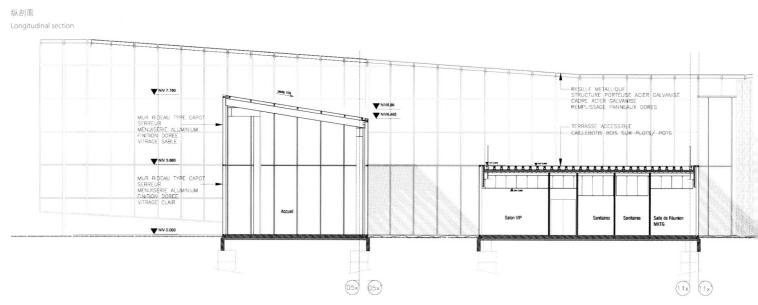

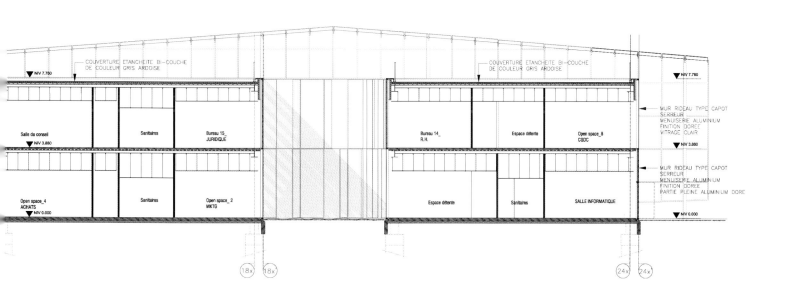

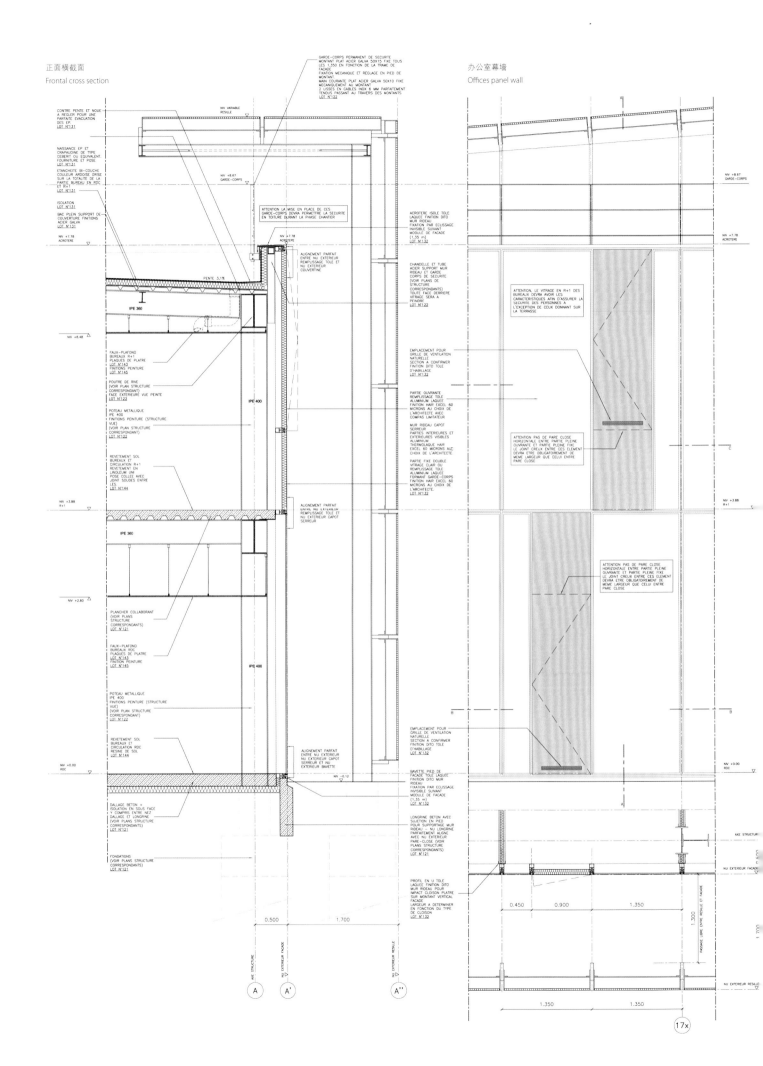

网状幕墙
Hairnet panel wall

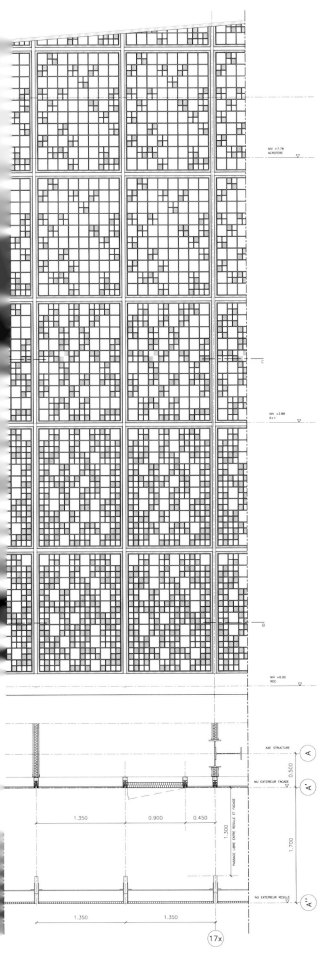

The new head office for the Piper Heidsieck and Charles Heidsieck House of Champagne faces onto the production unit, extending a wine-growing site and a presentation space. The same architectural style coalesces all the program elements and allows the new head office building to stand out as a contemporary yet unostentatious project reminiscent of an immaterial barn.

By raising its silhouette in the landscape, the building transforms the wine-growing site and becomes a beacon-like federating structure.

Inspired by the structural minimalism of winery buildings, the simplicity and efficiency of its delicate metal structure also suggests considerable elegance and sophistication.

The new head office is designed as a glass envelope enclosed within a grid that seems to dematerialize as it reaches upwards to the sky, providing an impression reminiscent of the effervescence of champagne bubbles. By day, the grid combines its golden reflections with the ever-changing colors of the sky in the Champagne region. By night, the building is subtly lit and the elevations become sources of diffused light.

Floating above the hard-surfaced esplanade, a spectacular canopy stretches out towards the landscaped car park to welcome visitors. The covered esplanade extends into the large reception hall where the light from the glazed roof is filtered by the metal grid and its golden panels. The overall impression is that of a stained glass window whose shards of light provide the interior space with an added colored sparkle.

The transparent heart of the building provides views of both the external landscape and the various interior work spaces formed from four pavilions separated by three patio gardens. The result is an overall sense of serenity emphasized by the important role played by nature.

The building combines the formality of an important house of champagne and the comfort of an environment favorable to human activities. By placing emphasis on use and an understanding of its landscaping role, the new Piper Heidsieck and Charles Heidsieck Champagnes head office succeeds in being both efficient and generous.

入口大厅
Entrance hall

Credits

Location: Reims, France
Completion Year: March 2008
Area: 2,000 m²
Architectural Team: JFA / jacques ferrier architectures
Architect: Jacques Ferrier
Project Manager: Stéphane Vigoureux
Team: Anna Sanna (Project Leader) Corentin Lespagnol (3D), David Tajchman, David Juhel, Elli Nebout
Interior Design: Pietro Ferruccio Laviani
Landscaper: Agence TER
Consultant Engineer: SNC Lavalin-Pingat Ingénierie S.A.S
Client: Champagnes Piper Heidsieck and Charles Heidsieck
Photography Credits: Jacques Ferrier Architectures/ photo Luc Boegly

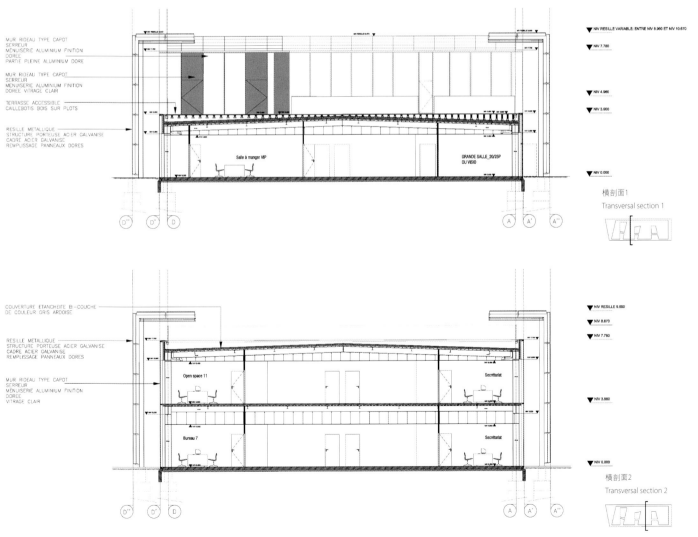

横剖面1
Transversal section 1

横剖面2
Transversal section 2

由金属与塑料丝网打造的112大楼外形像是一个白色的箱子……
The white box building with metal structure and plastic net....

112 Building in Reus

白色箱体——雷乌斯112大楼

| ACXT Architects |

雷乌斯112大楼是西班牙加泰罗尼亚地区新型紧急事件管理与服务体系的样板建筑,同时也是该国第一座获得LEED认证的公共设施。该建筑采用了全新的构造,将加泰罗尼亚所有负责管理紧急事件的机构都集中在一起。之前,这些机构(112呼叫中心、警局、消防站、公共卫生部门和民防单位)均分散在各地,各有各的电话号码,也没有共同的技术管理体系。将他们聚集在同一屋檐下,共享技术与工作流程,再把电话号码统一为112,可以更有效地协调管理各种紧急事件。

设计理念

项目整体布局清晰简洁,符合各种设计标准,同时主次分明:
① 场地的规模:在加泰罗尼亚地区所处的地理位置,与基础设施的关系以及人造景观;
② 社会与运作机构之间的关系(先前未显露出来):建筑外观,全新的应急管理系统,24小时提供服务;
③ 同一栋建筑中融合了各个应急管理机构:这座建筑内外均统一采用白

色，将所有机构汇聚在一起；
④ 各应急机构之间的关系：对于促进协作的公共空间的定义（以前，他们远程分享信息，如今却能畅通无阻地沟通）。

外形与功能

建筑从水平方向上被划分为三部分：服务区、公共区和操作区。服务区（停车场、更衣室、储藏室、休息区、问询台）的设计很符合场地独特的地形地貌。基座的顶部有一片景观区域，与建筑一层的公共区域（礼堂、新闻发布室和餐厅）相连。这一层位于一片橄榄树

立面图
Elevation

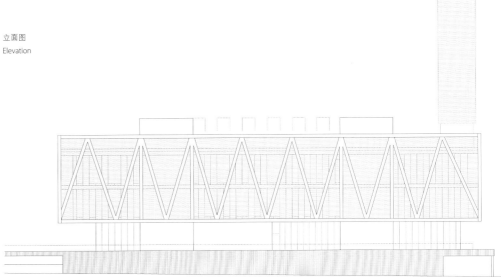

丛之中，不但能欣赏到周边地区的美景，还能通往楼上的操作区。

操作区的结构如同一只大箱子，由大型金属结构和塑料网构成，实现了双重设计目标：避免外立面直接接触阳光，同时外立面无窗式设计增强了建筑的地域特色。

这种金属结构除了能满足各个工作室的功能需求之外，还可以在未来灵活调整楼层的分布，凭借其坚固的外形和比较公正的白色，还能在新型应急管理模式下促进各个机构同心协力解决紧急事件。

为了加强各种操作之间的协作，工作区都围绕一个巨大的空间分布，这里自然光线充足，也照亮了操作区。

除此之外，垂直方向由四大核心结构组成：电信塔楼（也可以用于招待游客）、政府工作区、维修区以及工作人员的主要出入口。这个功能矩阵（水平层次和垂直方向）与建筑的结构与概念组成保持一致：安装有网状地板的一楼（公共设施基座）和作为工作区域的金属箱形结构由四大混凝土核心筒支撑着。

节能理念

这栋大楼不仅符合西班牙所有节能建筑条例规定，还获得LEED银级认证。大楼的设计采用高度节能的外立面及服务设施。同时采用高效管理机制，鼓励公共交通与汽车合用。

与普通建筑相比，这栋大楼的建造节水50%，能源消耗减少34%。

一层平面图
First floor plan

二层平面图
Second floor plan

操作区
Operational level

走廊临中庭方向的玻璃挡板选用白色包边，形成与外立面之间的自然过渡
The glass rails along the passage facing the central atrium is white framed, in harmonious accordance with façade

剖面图
Section

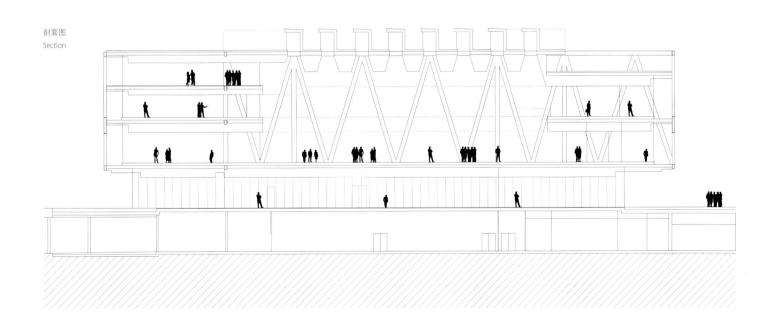

会议室临中庭方向的墙壁同样采用白色边框的透明玻璃建造，营造出无限通透的空间感
The white framed glass walls of the conference room add a sense of space brightness

The new 112 Building in Reus is the model for the new emergencies management and service system in Catalonia, and the first public facility in the country to have a LEED certification.

It is a new architectonic typology that brings together all the bodies in charge of managing emergencies in Catalonia.

Before, these bodies were scattered around the territory (the 112 call center, the police, firemen, public health and civil defense), had different phone numbers and didn't share technological frameworks.

Their gathering under a same roof, where they share technology and processes, and the substitution of all lines for the single emergencies number 112, will result in a more efficient and better coordinated management of emergencies.

Concept

The complex follows the rationale using an organizational system based on clear and concise project tools which help standardize its heterogeneous demands and highlight what we consider important:

• The scale of the place: location within the territory, relation with infrastructures, artificial landscape.

• The relationship (previously hidden) between society and operatives: exterior view of the building, showing the new emergency

行动协调中心
Operational coordination center

management system, day and night presence.

• The coming together of the operative bodies in one building: a single structure holds them all, unified by the color white, transversal to them all.

• The relation between operative bodies: definition of a common space that activates synergies (before they shared information from a distance; now they communicate).

Shape and Function

The building is horizontally divided into three layers: services plinth, public area and operational levels.

The services plinth (parking lot, changing rooms, stockrooms, resting areas, building services) becomes the element that adapts to the peculiarities of the plot (topography and shape). Its roof offers back to the environment a landscaped area, which meets with the public part of the building (auditorium, press room, restaurant) on the first floor. This floor, situated above the olive-tree fields, grants excellent views over the territory, leading to the next floors: the operative box.

The operative box is formed by a large metal structure and a plastic mesh that achieves a double objective: avoiding direct solar radiation on the façade, on all sides, and cancelling out the idea of openings in the façade, thus increasing the territorial aspect of the building.

The metal structure, apart from solving the functional needs of the operational rooms, also allows flexibility to be incorporated for future floor distributions and promotes an image of unity of the operational bodies integrated in the new emergencies management model, through its solidity and its white color, neutral to the uniforms of the operational bodies.

To boost the coordination and the synergies amongst the operators, the operations rooms are arranged around

停车场，顶棚种植了植被
The parking lot with a plants covered roof

Credits

Location: Reus, Tarragona, Spain
Completion Year: 2010
Total Area: 14,985 m²
Project: Emergency Call Management Center—112 Building
Client: Ministry of Home Affairs. Government of Catalonia
Architect: Marco Suárez
Assistants: Elida Mosquera (P+O), Jonathan García (O), Sorana Radulescu (P), Roberto Molinos (P), Mireia Admetller (P), Claudia Carrasco (P), Alexandre Borras (P)
Photographer: Adrià Goula

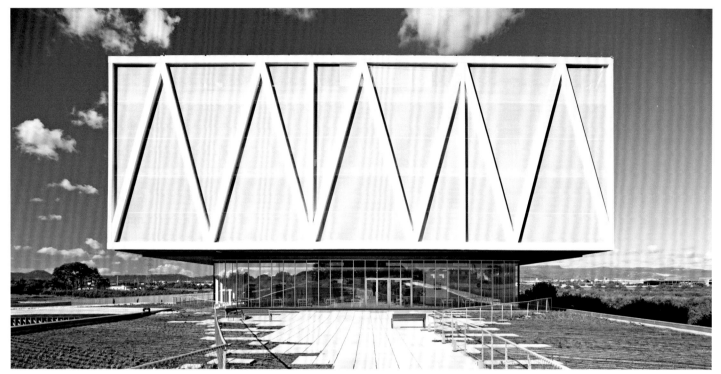

上层白色箱体结构在周边美景的映衬下别有特色
The box structure above promotes a grand beauty among the landscape

a large space that also allows natural diffuse sunlight to the inside of the operational box.

Otherwise, vertical communications are defined by four nuclei: the telecommunications tower nucleus—which is also useful when receiving visits—the authorities' one, the maintenance one and the main access for workers one.

This functional matrix (horizontal layers and vertical communications) coincides with the structural and conceptual configuration of the building: a ground floor with a reticular flooring (services plinth) and a metal structure box, that forms the operative area and is supported by the four concrete nuclei.

Functional and Physical Safety

In order to protect operations, the building has high security measures, both exterior and interior. The need to work round the clock, 24 hours a day, 365 days a year, calls for the main building services (electricity, telecommunications, air conditioning) to be redundant and for the building to be able to be self-sufficient for at least 5 days, should the supply fail.

Energy Efficiency

The building, apart from satisfying all Spanish energy efficiency regulations, has obtained a "LEED Silver Certification". This not only implies the achievement of high energy efficiency values (façade, services) but also, that a more demanding building process has been carried out. It also implies the use of mechanisms for building management and the workforce, after construction, for example encouraging the use of public transport or car pooling, among others.

It is thought that the systems planned will save 50% of water and 34% of energy consumption, compared with that of a normal building. One of the highlighted mechanisms is the use of dissipated energy from the Data Processing Center for heating the water that the building needs or the usage of a geothermal system to heat/cool the changing rooms.

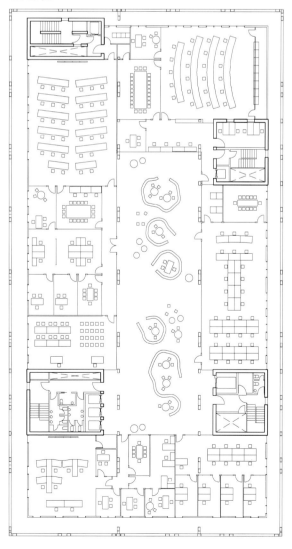

三层平面图
Third floor plan

别墅的两部分通过两层楼高的起居室连接起来,起居室的全玻璃外立面让室内显得更加宽敞
The two low volumes of the house are connected by the two-storey space of the living area, completely glazed to underscore the size of the interior

Villa T | T别墅 |

| Architrend Architecture |

拉古萨城地处伊卜利奥高原,这里曾经到处是干燥的石灰墙,T别墅就在城郊的乡村。

在这里,一幢幢高楼拔地而起,逐渐改变着这里的原貌。资源开采和畜牧相关产业曾经在这里发展迅猛,而这些产业活动曾经在农村、农场上以石头建成的小屋(饲养员的住处、马厩及奶制品厂的原奶加工地)内进行。别墅的选址维持了当地原貌,因此在那里可以一览伊卜利奥高原的迷人风光。

项目的基本要求是建立和当地美景、风光之间的视觉联系,当然这一切的设计都无需参照和模仿周边农田民舍的建筑结构。现代化的元素将别墅其与周边的景观建筑有机结合,实现了两者之间的自然过渡。

东立面
East elevation

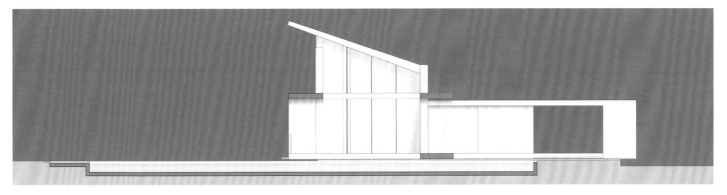

南立面
South elevation

剖面图
Section

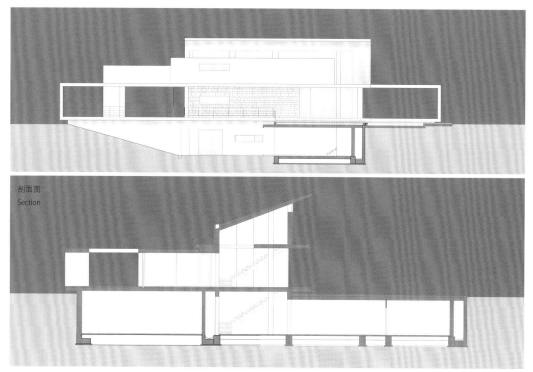

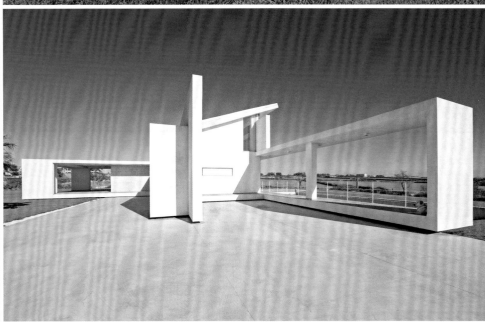

一楼平面图
Ground floor

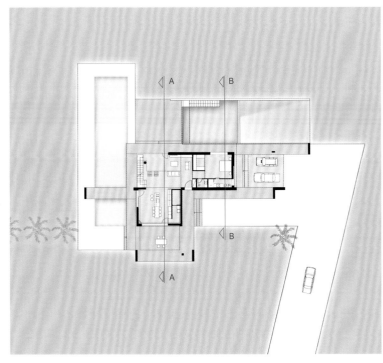

而别墅与环境之间的联系则取决于线性框架的设计，让欣赏周边景致的视野更为开阔。

别墅的北侧被石墙围合，而南侧和东侧则通过大型落地窗与花园相通。倾斜的屋顶由轻质玻璃设计而成。别墅的基座被抬高，脱离地表而建，显得更加轻盈，而移动墙、非对称外墙以及环形墙则使房内动感倍增。

别墅有三层，底楼被分为双层起居室和起于厨房的餐厅，厨房向外延伸至室外区域，该区域四周分别是主卧、卫浴套间以及配有壁橱的更衣室。

起居室的窗户朝向已建的农场及周边区域。楼梯以钢结构和木制台阶为主，从起居室向上延伸到阁楼的主卧和书房，向下连接地下起居室，从那里可以看到地下的天井。该楼层上还有两间其他房间和相关的设施，从这里可以望见外面的庭院。

大楼立面大部分被粉刷成白色，这可以烘托多样化的建筑元素，并和周边的地中海风光相辉映。

外部的地板采用灰色水泥铺置，并且对边角进行了修整，使表面平整光滑，而屋内则使用厚厚的橡木地板。而底楼的卫浴套间则铺设了沥青石材（表面涂有沥青的当地石料，呈独特的灰暗色调）；另一处卫浴套间的墙壁采用树脂材料建造。

设悬式木质台阶的楼梯沿着客厅的玻璃立面通向阁楼
View of the mono-beam staircase with cantilevered wooden steps along the glazing of the living area, reaching the loft with the studio

The villa is located in a rural area on the outskirts of the city of Ragusa, Ibleo Plateau characterized by a network of dry stone walls in limestone.
The area has undergone a gradual process of building that has changed its original features, especially related to the exploitation production and cattle breeding, and these activities were conducted in rural villages, farms, consisting of a complex of buildings in stone (home of the breeder, stables, places of transformation of raw milk in dairy products), around a courtyard called beam. The project site partially maintains these characteristics and therefore has a very impressive view of the Plateau Ibleo.
The primary requirement of the project was therefore to establish a direct contact with the beauty of the area, with its landscape, and of course all this was done without any camouflage and reference architecture of rural farms, but also building, with contemporary elements, a dialectical relationship of mutual exploitation where the architecture created by the landscape and established a fruitful dialogue with it.
The requested relationship with the environment is first entrusted to the implementation of three linear frames that highlight the views of the landscape. One-sided enclosed with stone walls facing north is a counterpoint to the south facade and the east, opening to the garden through the large windows. The sloping roof is made from light deadlift glass on the outer walls. The base of the house is raised off the ground, creating a shadow line and a consequent feeling of lightness, while slips, asymmetries and the articulation of the walls create a

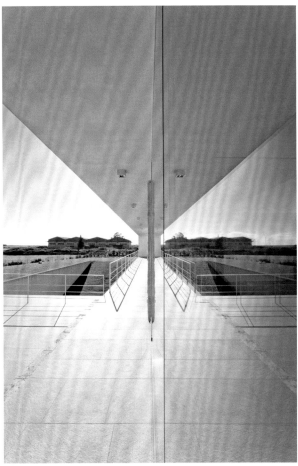

客厅的玻璃立面
View of the glass facade of the living area

refined idea of movement.

The villa has three levels, the main ground floor is divided by a double height living room, dining room from the kitchen which extends to an outdoor area protected from the master bedroom and related services, bathroom and dressing room with wardrobes.

The windows of the living room, angle without metal profiles, and oriented towards a view on an existing farm and the surrounding area. From the living room a staircase in steel and wooden steps cantilevered rooms to loft with bedroom and study area, and falls into a large combined family living overlooking a patio sunk in the ground. At this level were located two other rooms and related services, with a view of the courtyard.

The elevations of the building are almost totally plastered in white, a color that highlights the various architectural elements and link with the Mediterranean context.

The external floors are porcelain slip-gray cement placed to run without escape with rectified edges to make the surface smooth and internally flooring planks of oak pre-finished with a brushed surface has been used. The bathrooms on the ground floor have coating strips of asphalt stone (a local limestone with the presence of asphaltic bitumen that makes them take the typical dark color). The other bathrooms have a resin finish on the walls.

从厨房到客厅,同样的橡木地板造成了视觉上的连续性
The perspective from kitchen toward the living area creates a visual axis open to the landscape, underline by the continuity of the prefinished oak floors

Credits

Location: Ragusa, Sicily, Italy
Materials: Steel, Glass, Stone
Designers: Architrend Architecture, Gaetano Manganello & Carmelo Tumino
Contributors: Patrizia Anfuso, Fernando Cutuli
Photographer: Umberto Agnello

这栋别墅的主人喜欢石雕,因此建造了这个65平方米大小的作坊
A 65-square-meter workshop provides a space for the homeowner's passion for stone sculpture

A Modern Villa | Lehrer Architects |
浑然一体——南加州林间别墅

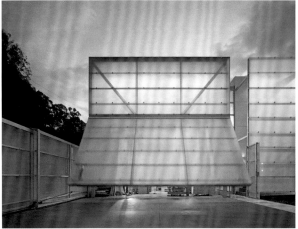

作坊的玻璃门可以开合自如
Glass door opens for workshop use and when closed keep a graphic-streamlined profile.

显然，Lehrer建筑师事务所的精心设计给这座既时新又舒适的传统别墅增添了无限现代化特色。建成的别墅几乎涵盖了现代别墅的所有元素，它弥合了室内室外的分界线，兼及透明和坚固的质感，通过自然光照和玻璃外观使建筑主体透明化。别墅内的水平和垂直景观在楼内任意观景点都能欣赏。别墅的钢框架结构被镀上光滑的白色灰泥，并安装了滑动玻璃墙和可移动天窗，既宏伟壮观，又与大自然融为一体，其中，移动玻璃墙面向花园和周围的林木，而可移动天窗则使建筑顶部在视觉上具有通往天空的通透感。

为使室内外设计过渡自然，保持内部空间流畅，在别墅规划中，设计师尽量展现别墅在水平与垂直两个方向上的紧密契合：水平方向上，独栋别墅围绕花园露台而建，一个隐秘的私人花园延伸至底楼的起居空间；而垂直桁架则将大楼的各部分连接和组织为完整的一体。靠近楼幢桁架的别墅空间形成墙内的窗口，可纵览别墅周边及花园的全景。别墅底楼被76.2厘米基准面和玻璃表面包裹，营造出别墅的漂浮感，并与周边的景致相互交织。上方的天窗更增强了与景观的融合。别墅内的每一间房都安装了各式落地玻璃门，包括旋转门、推拉门与内置滑动门，充分将室内外景致融合

可移动式玻璃幕墙面朝花园与周边的树林,视野开阔
Expanses of sliding glass walls open out to the garden and surrounding woods

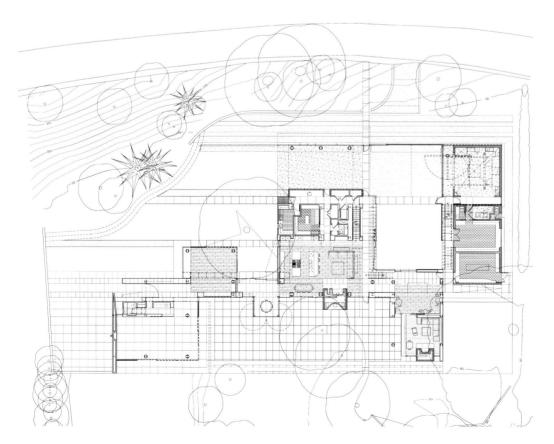

为一体。白色灰泥、钢铁、白色的原木地板以及玻璃架构增添了别墅的光辉色彩。

起居室位于底楼,占地83.6平方米,楼内的玻璃外墙四面开放时,起居室俨然一座巨大的室外楼宇。楼面的角落处和周边环境浑然天成,使其从根本上成为了一处有顶的"户外空间"。其中,一处用白玛瑙设计而成的小酒吧尤为惹眼。

通透的厨房与家庭房融为一体,厨房采用亚麻色的原木木料搭配不锈钢装置。方体早餐室位于日式花园对面,散发出丝丝禅意。主楼梯形如细工家具,落地式玻璃栏杆从底楼延伸至顶层,而木阶梯则跨越三层楼。

在主卧室内,房间自身演变成了一处61米长的床头板,那里可以畅通无阻地通往室外的平台,这也是世外桃源般的放松去处。宽敞的顶级套间延伸至一处露天的浴缸,坐在浴缸内可以俯瞰花园。室内、室外的淋浴设施顶部开放,凸显了户外设计风格,镀膜不透明玻璃保证了私密性。顶级的卫浴套间通往雕像工作室和健身馆,套间内的无光玻璃镜子光滑细腻,水槽悬于玛瑙底座上方。

别墅三楼巨大的玻璃浮桥连接着助手办公室、一处巨大的复合式甲板以及健身馆。而可移动天窗则使别墅的透明度倍增。

纵观整个别墅,Lehrer建筑师事务所对细节的专注在与总承包商Horizon的合作中体现得淋漓尽致。比如,由15.2厘米厚的大理石构成的工作台面,选用来自意大利的10余种大理石、花岗岩以及石材精制而成。设计师为了达到统一的外观效果,对高于柜台高度的厚板进行切割,对可拓建筑部分进行倾斜,并沿着厚板前沿做出向下"折叠"的观感。"对高尚、高雅的追求使这座别墅集简约、简单及复杂建筑结构于一体——简直是一首由大小主题构成的交响曲,"建筑师Michael·B·Lehrer如是说道。

可持续性策略:这是一座零能耗的住宅,住宅内采用被动系统,阳光充足,空气清新。设计师对窗户的位置与外形仔细进行斟酌,以求达到最大化的自然光照和最小化的热量吸收。南面的玻璃外墙贴有瓷片,夏天可阻挡炎热,同时,悬垂结构可用于遮阴。由于别墅周边微风徐徐,所以无需空调设备。取暖很重要,所以地面安装了地热系统,该系统由位于地表下方的管道系统组成,可以压泵取水从而散热。房屋顶部安装太阳能电板集热,还可以利用太阳能来烧水,从而最大化的使用可再生能源。屋顶的太阳能电板保证了居民100%的能耗需求,包括业主的特斯拉电动车和雕塑工作室在内,太阳能电板可以满足98%的用电需求。

钢筋结构表面涂上了光滑的白色灰泥,干净清爽
The steel-frame structure is clad in smooth white stucco.

白天,将玻璃落地门打开,整个起居室就变成了一个户外的亭子
The living room becomes an open-air pavilion when floor-to-ceiling glass doors are open

周边自然景观在白色建筑物的映衬下更加青翠欲滴
The natural surroundings shine when framed by the home's white background

别墅的朝向与地理位置取决于四棵参天大树
The house's orientation and footprint were determined by the property's four significant trees

一楼平面图
First floor plan

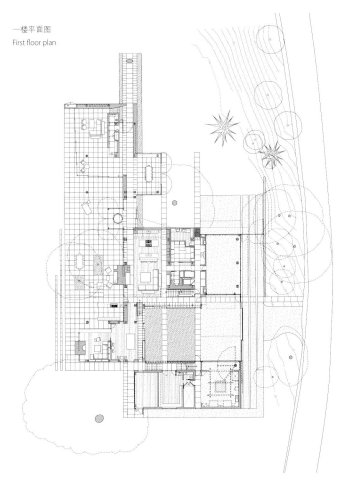

Lehrer Architects conceived the project as a timeless and comfortably elegant agrarian villa in a distinctly Modern tradition. The end result is a veritable "Encyclopedia of the Modern Villa" that blurs all boundaries between indoor and outdoor spaces, is a seamless play of transparencies and solids, and dematerializes structural mass through abundant natural light and glass. Horizontal and vertical view slots are integrated throughout the house, as are sightlines to the entire property from every vantage point. Spectacular and at one with nature, the steel-frame structure is clad in smooth white stucco and expanses of sliding glass walls that literally open out to the garden and surrounding woods, and walkable skylights that visually open up to the sky.

Driven by spatial clarity and informed by indoor/outdoor unities, the design exhibits intense horizontal/vertical rigor, with the grand-scale single-family residence organized around two planes: the horizontal plinth of the garden patio that extends the private garden into the living spaces of the ground floor, and the vertical spine that connects and organizes the various parts of the building program. Spaces that abut and traverse the spine create openings within the wall, framing views and vistas of the site and garden.

A 30-inch datum/glass wrap at the ground floor level gives the feeling that the house is floating, enhancing the graceful blending with atmosphere and surroundings that inspires the design. Clerestory windows above reinforce this sensibility. Almost every room in the house has floor-to-ceiling glass doors—pivoting, swinging, or pocketed—for fully immersive indoor/outdoor living. A minimal palette of white stucco, steel, pale wood, and glass further orchestrates the lightness that defines the house.

On the ground floor, the sweeping 900-square-foot living room becomes a

楼梯被打造得如细木家具般精致
The main stairs are built like cabinetry

过道顶部的天窗为室内增添了一丝通透
Skylights along the hallway add another layer of transparency

二层平面图
Second floor plan

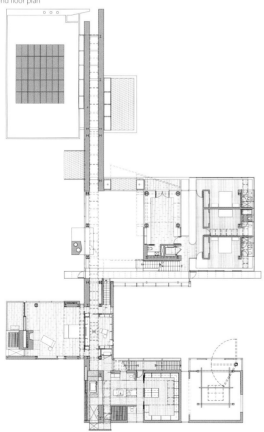

dramatic outdoor pavilion when all the glass walls are opened. Corners are freed for seamless flow, and the environment becomes, in essence, a covered outdoor space. Notable design elements include a wet bar in white onyx.

The airy kitchen, which merges into the family room, features blonde wood islands with clean stainless steel fixtures. A breakfast room exists as a Zen cube set into the open exterior space, adjacent to a Japanese garden.

The main stairs are built like cabinetry—with full-height glass railings seemingly inserted into the design—and climb from the ground level to the upper floors. Articulation of the stairs in wood in the three-story volume echoes the ascent from exterior views.

In the master bedroom, the house itself becomes a 200-foot-long headboard, and the uninterrupted flow of space into the outdoors beyond creates an idyllic retreat. The spacious master suite flows into an open-air soaking tub overlooking the garden. Indoor and outdoor showers—with fogged glass for private areas that is graduated so there's still a view on top—extend the al fresco experience. The master bathroom—which provides direct access to a sculpture studio and gym—features mirrors with etched glass for a silken finish, and sinks cantilevered on onyx bases.

On the third floor, a dramatic glass bridge connects an assistant's office, a large recycled composite deck, and gym. The walkable skylight adds another level of transparency. Geometry is broken at this height to pay homage to the towering Sycamore tree, with a mesh parapet and railing creating a void zone.

Throughout, Lehrer Architects' signature attention to detail is brought out by working closely with Horizon General Contractors. For example, each of the

家庭房与厨房格调一致，微微泛着金色木头的光泽
Blonde wood unites the open family room and kitchen

more than 10 types of marble, granite, and stone were personally selected by the owners in Italy for countertops that appear to be made of six-inch slabs of marble. However, by cutting the slab longer than the counter depth, beveling the extension, and "folding" it down along the front edge, the patterning maintains for a continuous look. "Detailing is high-toned and elegant to assure that this Modern Villa sings coherently as one singular, simple, complex composition—a symphony with major and minor motifs," says architect Michael B. Lehrer, FAIA.

Sustainability: This is an (electrical) energy net-zero residence that is flooded with natural light and fresh air through mostly passive systems. Window placement and type were carefully considered in terms of balancing maximum natural light with minimal solar gains. South-facing glass walls are accented with ceramic frit to block heat on warm days, and structural overhangs were designed to increase shade. Set within a breezy climate, the house needed no artificial air conditioning. Heating was more of a concern, so the floor hosts a radiant system comprising tubing set beneath the surface and pumped with hot water. The structure is topped with photovoltaic panels for electric power, and solar absorbers for heating water, maximizing the allowable energy production levels as determined by LA Department of Water and Power. The roof-top photovoltaic panels produce 100+ percent of the residence's power requirements. With the owner's Tesla electric car and sculpture studio included, the photovoltaic's produce 98% of the total electricity requirements.

厨房采用光亮的金色木头与不锈钢打造，现代而简约
Sleek blonde wood cabinetry and stainless steel create a modern kitchen

三层平面图
Third floor plan

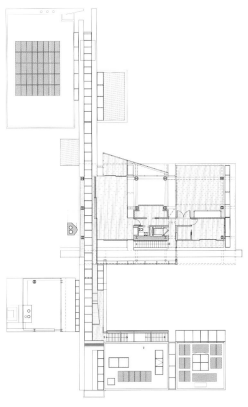

玻璃通道内光线充足
The glass walkway lets light penetrate into the space.

卫浴间洗手池台面采用缟玛瑙石打造，古典味十足
Master bath sinks are cantilevered on onyx bases

精细的布局
Meticulous attention to detail is seen throughout

Credits

Location: Southern California, USA
Building Area: 1,231 m²; Three Floors
Site Area: 5,666 m²
Lighting Designers: Lighting Design Alliance, John Brubaker Architectural Lighting Consultants
MEP Engineer: IBE Consulting Engineers
Structural Engineer: Reiss Brown Ekmekji
Civil Engineer: VCA Engineers
Soils Engineer: Geocon
Architects: Lehrer Architects
Principal-in-Charge of Design: Michael B. Lehrer
Project Architect: Robin Sakahara
Project Designer: Erik Alden
Project Team: Christian Arndt, Steve Deyer, Dongwoo Hyun, Chika Ito, Tim Jordan, Lucy Kelly, Chris Mundweil
Interior Designer: Unique Custom Interiors
Principal: JoAnne Brosnaham1
Photos: Benny Chan/Fotoworks

ROC Rijn IJssel
ROC社区教育中心

|LIAG|

为了提升新学校的形象,体现普莱斯克哈福社区及社区大学的发展,这栋透明建筑物应运而生。建筑的结构和内部的活动在室外都清晰可见。建筑顶端的悬挑,巨大的玻璃立面和缠绕在四周的楼板边缘,赋予了建筑雄伟大气的外观。设计师将原有建筑中的红地毯作为主要元素贯穿整栋建筑:宽阔通达的楼梯和露台。在楼梯间设有半封闭式透明窗户。楼梯赋予了建筑透明的特性,同时也连接了建筑的每一个角落。

大楼室内空间布局灵活,开放的特性保证了学习环境的舒适,并能适应目前及未来建筑调整需求。建筑的外形是紧凑的,最大限度地减少了外立面面积,可以保留更多的能量。项目自身的功能要求提供多样性的教室以满足众多的教学科目要求。设计师选用柱状结构分离空间,同时对立柱之间的距离与高度仔细斟酌,保证空间布局灵活,历久弥新。

一层平面图
First floor plan

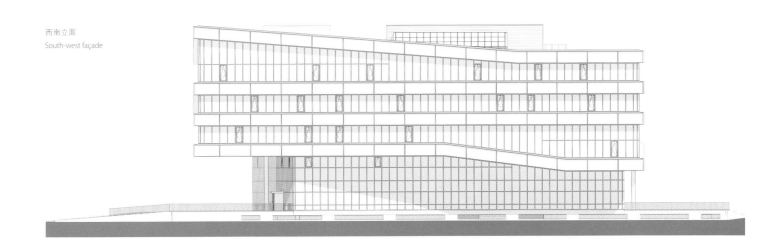

西南立面
South-west façade

剖面图1
Section A

In order to elevate the new school building to symbolize the redevelopment of both the neighborhood and the regional community college, a highly transparent building was created with practice rooms oriented to the surrounding residential area. The structure of the building and activities going on inside can be observed readily from the outside. The generous overhangs on the top floor, the interplay of lines around the edges of the floor and the huge glass surfaces lend the building a powerful, striking appearance. The red carpet

剖面图2
Section B

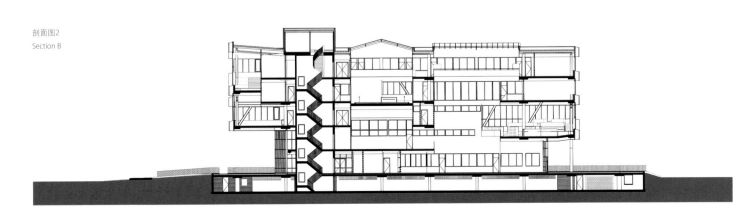

长而宽阔的楼梯从二楼通往三楼
The long wide stairs connecting the second floor to the third floor

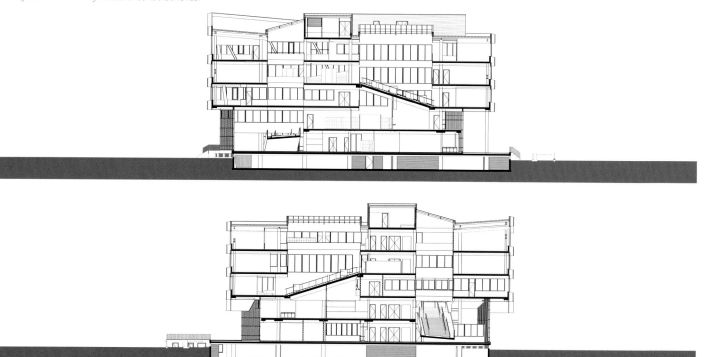

Bright interior atrium

from the original educational vision was used in the final version of the concept as an organizational element throughout the entire building: long, wide stairs and terraces dissect the floors. Alcoves were added to the stairs; semi-closed, but transparent. The stairs provide access to all of the teaching areas, which adds emphasis to the transparent character of the building.

This is a flexible building, with an open character that offers a secure and comfortable learning environment, and meets both current and future educational needs. The shape of the building is compact, which minimizes the face of the outside wall and consequently retains more energy. The wide range of study programs required considerable diversity in terms of rooms. However, demand could easily change in the future, which is why flexibility was a non-negotiable requirement. LIAG responded by choosing a column structure with column distances and adjustable heights in order to allow for various layouts. The column structure enables flexible use, thus ensuring a long lifespan and occupancy.

Credits

Location: Arnhem, the Netherlands
Completion Year: 2008
Area: 14,518 m²
Subject: New Build Vocational Secondary Education
Functions: Education Building with Theory and Practice Rooms, Workshops, Offices and Examination Rooms and Parking
Photographers: Ben Vulkers, Iemke Ruige
Client: Board Rijn Ijssel

带屋顶花园的新会议中心
The new Congress Center with roof garden

Hangzhou Congress Center

杭州新市政大楼

| Peter Ruge Architekten |

经过多年的设计和建造，杭州新市政大楼终于落成。大楼的立面设计是由Peter Ruge建筑师事务所（前身为Pysall Ruge建筑师事务所）与王晓松教授精诚协作完成的。
这座市政中心大楼位于钱塘江附近，距市中心也不远，它将是杭州市新的大型商务及行政区的中心。市政大楼包括6座办公楼，

屋顶花园细部
Detail of roof garden

1. 屋顶花园 Top light
2. 屋顶天窗 Roof garden
3. 钢结构 Steel construction(lamellas)
4. 外立面 Facade construction

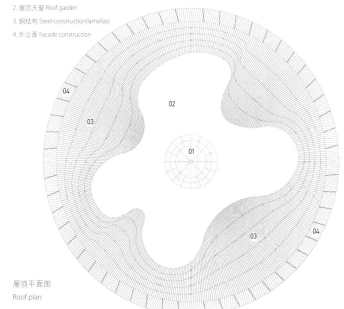

屋顶平面图
Roof plan

它们由位于高处的圆形天桥连接在一起。高层楼的侧面还有一些多功能的单层建筑，从四个方向都有到那里的入口。而杭州市政府则位于这六座高楼的中心，象征着宝石。

外立面既要起到支撑作用，又要让市政大楼成为一座融入当地传统元素的现代建筑。

浙江省以茶叶生产而远近闻名，为了将这一地域特色体现在大楼上，幕墙的设计采取了茶叶种植田地重叠向上和网格型构造。最后形成的是一个多层的结构，完全体现了建筑的可塑性。从远处看大楼的幕墙是一个坚固的整体，但是走近了就会看到它的网格结构。

楼顶的设计理念就是让它成为大楼的第五立面，这样就会让市政中心大楼看起来更有当地色彩——整个市政大楼看起来宛若一朵荷花，从六座高层大楼的顶部几层俯瞰，这样的效果就更加明显了。幕墙结构一直延伸到市政大楼的楼顶，将大楼的一部分包围起来。不同长度和固定高度钢筋的使用让楼顶呈波浪形，并让未被幕墙结构包围的部分从高空看也是一朵荷花的模样。这些未被包围的部分上面栽种了许多植物，成为大楼的绿色景观。

"我们的目标就是让这座大楼融合并表现该地区所有的自然特色，这样，当地人才会有身为杭州人的归属感。"建筑师Peter Ruge说。

整个市政大楼建筑群
The entire new City Administration of Hangzhou

会议中心
Congress Center

立面图
Elevation

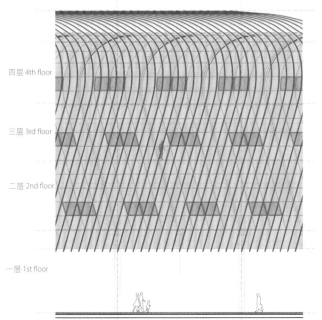

四层 4th floor
三层 3rd floor
二层 2nd floor
一层 1st floor

After perennial design and construction phase the congress center of the new city administration of Hangzhou, China is completed. The concept and design of the facade was made by the architect Peter Ruge and his team in collaboration with Prof. Wang Xiaosong. The new building ensemble is situated close to huge Qiantang River not far from the city center. It will be the focus building of the new large business and administration district of the city. The new fascinating complex consists of six office high-rise buildings arranged in a circle and connects in the upper floors through a circular bridge building. The high-rise buildings are flanked with flat multi-functional buildings including four main entrances from all directions. As the new central form of the main administration building of the City of Hangzhou the Congress Center resembles a large precious stone.

The façade design should support on one hand the unique modern architecture of the building ensemble but on the other hand it should take up typical local or traditional aspects of the region also.

剖面图
Section

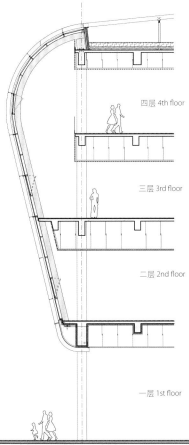

四层 4th floor
三层 3rd floor
二层 2nd floor
一层 1st floor

Zhejiang Province is known for its tea-producing region. To express the building's regional characteristics, design of the façade is based on the superimposed configurations of the tea cultivation pathways and the planting nets. As a result, the building is enveloped by a multi-layered fabric, giving it a true architectural plasticity. Seen from a distance, the façade appears like a rigid volume, but dissolves into a network of structures and levels as you come closer.

The main idea for the design of the roof was to use it as the fifth façade of the building to set up a strong and typical local image in the shape of a lotus blossom, which you can see from all upper floors of the surrounding high-rise buildings. The façade structure would be extended unto the roof of the congress center to cover up it partly. Through the different lengths and fixed height of the steel beams the structure is waved and forms the abstract blossom of lotus in the center of the roof. This part isn't covered and is designed and planted as a green landscape.

"Our aim is to combine and express all the regional natural features within the Center, so that the local people will be able to identify themselves with the City of Hangzhou," says Peter Ruge.

Credits

Location: Hangzhou, China
Completion Year: 2011
Architects: Pysall Ruge Architekten~Peter Ruge, Matthias Matschewski, Nicole Kubath
Project Partner: DBH Stadtplanungs GmbH - Prof. Wang Xiaosong, José Maria Cantalapiedra Alonso
Façade's Consultant: Schlaich Bergermann und Partner
Photograph: Jan Siefke

磨砂玻璃肋板如同轻盈的纸片悬浮在立面结构上
Like light thin sheets of paper, frosted glass fins float from the structure

Arkansas Studies Institute
阿肯色州研究院

| Polk Stanley Wilcox Architects |

项目描述

阿肯色州研究院位于一座大型图书馆和一所州立大学之间，这里收藏着1 000万份历史文献，以及包括威廉·杰斐逊·克林顿总统在内的七任阿肯色州州长的手稿。

研究院位于繁华的娱乐区内，靠近克林顿图书馆，设计师以高科技呈现富有表现力的新建筑形式，将之与那些19世纪80年代以及20世纪初被忽视的建筑结合起来，从而创造出了一条通往图书馆和阿肯色州"历史见证者"的标志性人行通道。

设计的思想主要基于"书本"——储存信息的实体容器，而翻动的页面则生动形象地勾勒了大楼的构造。大楼外立面上的多层曲形玻璃墙好像一本打开的"书本"的"页面"，访客可以从这些"历史"页面中穿行，从新时代走进古老的纪元。狭小的中庭将新楼隔开，使古建筑完好无损，大楼也因此更为狭长，光线可以洒进各个楼面——这是一项关键的可持续策略。中庭的玻璃扶手上挂着100幅历史悠久的画作，充分表明建筑能够而且应该积极地传达历史故事。画廊、咖啡馆、博物馆和会议室等公共空间更显临街店铺的熙熙攘攘，建于1914年的历史建筑二楼则全被研究大厅占据。中庭的玻璃墙上展示着文件，似乎在告诉人们，知识是触手可及的。一座吊桥将新旧两界、喧闹宁静、今世往昔连接了起来。

建筑以层叠式结构逐渐向后弯曲，从而避开场地上原本生长着的一棵树
The design purposefully curves back in layers to avoid the site's only existing tree

垂直的磨砂玻璃肋板看上去如同翻动的时间书页
Vertical frosted glass fins visually move, like flipping through the pages of time

位于西侧立面的磨砂玻璃肋板可以控制光照，并展现阿肯色州的历史面貌，犹如历史的书签。

封面

从研究院内存放的文字媒介中获取灵感，入口设计得如同书的封面一般，和大楼分隔开来，如同大楼的第二道墙壁，阻挡了从西面进入中庭的光照和热度。铜制材料经久耐用，类似于皮制的封面一般，建筑弯角处的黏合物都显露在外。书页式的设计在玻璃肋板，楼梯和街景中都可以看到。

构想

"书页"的建筑构想涵盖了整座大楼和项目场地，形成了一条始于克林顿大街、穿过研究院的走廊。

这里的街巷和马路颇具意大利市集的气息，凸起的石制减速带和粗糙的马路铺面可有效延缓交通事故的发生。

书页

大楼旁悬着的磨砂玻璃墙如同书本轻薄的纸张。档案储存室和大楼底座全部用阿肯色当地的沙石制成，象征着内部史料的重要性。

软楼层的美术馆向三个方向打开，通过各个方向的入口让建筑更加有生机。玻璃板可以折叠和重合，既能够保护场地上的唯一一棵大树，也同时展现其书页的形态，留给人无限遐想。

学习场所

主研究大厅位于移动建于1914年的古老建筑内，这座大楼曾经是仓库。中

大楼的新入口犹如书的封面
The new entrance acts as an abstract book cover

一层平面图
Level one plan

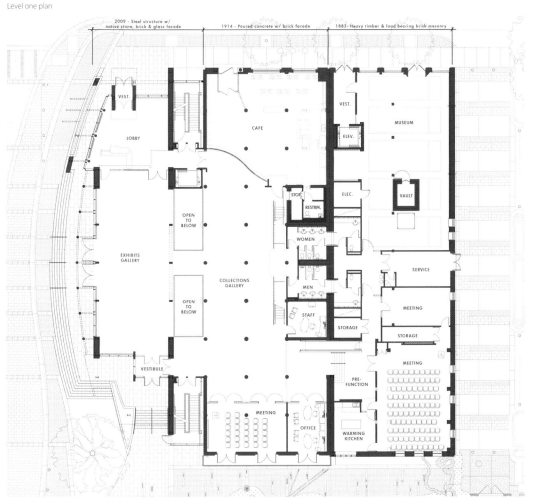

庭是安全规划的一部分，图书馆工作人员可以穿过吊桥到达文献室。

历史的展示

入口的中庭模糊了室内外的分界线，大楼和场地的界限。铺设砂岩的广场延伸进了中庭之内，成为档案文献室的底座。

走进大楼，映入眼帘的首先是美术馆、电梯走廊上的相框以及保存的文件。

历史的影子

历史是一种记忆，是往昔的回声，是过去的影子……然而历史也是鲜活而灵动的。

阿肯色州研究院使历史、研究、行人和丰厚的街景交汇一处，它弥合了城市肌理中的缺陷，向公共社会宣传环保理念，也成了知识的灯塔。

Galleries bridge time from new building to old, rotating to permanent collections

Program Statement

The Arkansas Studies Institute, a unique partnership between a metropolitan Library and a state university, is a repository for 10 million historic documents and the papers of seven Arkansas Governors, including President William Jefferson Clinton.

Located in a thriving entertainment district near the Clinton Library, the design combines significant, but neglected buildings from the 1880s and 1910s with a new technologically expressive archive addition, creating a pedestrian focused, iconic gateway to the library campus and the public face of Arkansas history. The design philosophy is based literally on the book—a physical container of information, with pages flowing into a site sensitive, physical narrative of the building's function. Multiple curving glass walls of the new main façade represent pages of an open book, where patrons literally walk through the pages of history, from new to historic spaces. A thin atrium pulls the new structure

away to protect the old, stretching the building's length and flooding all levels with light—a key sustainable strategy. 100 historic images in the atrium's glass handrails signify that architecture can and should actively engage in storytelling. Public Spaces—galleries, a café, a museum, and meeting rooms—enliven streetscape storefronts, while the great research hall encompasses the entire second floor of the 1914 building. Document storage displayed through the atrium's glass walls expresses that knowledge is within reach. Suspended bridges span the gap between new and old, open and secure, today and yesterday. The west façade's frosted glass fins control sun exposure while displaying historic

二层平面图
Level two plan

入口的中庭模糊了室内与室外、建筑与场地的界限
The entrance atrium blurs the edges between exterior and interior, building and site

faces of Arkansas life, like large book marks in time.

The Cover

Taking cues from the written medium for which the Institute was created, the entrance acts as an abstract book cover, pulled away from the building as a double wall, defusing the western sun's light and heat in the atrium beyond. The copper ages with use, not unlike a leather cover, with its binding exposed to the intersection. The flowing pages continue into the glass fins, steps, and streetscape as well.

The Idea

The conceptual idea, the pages of an open book, framed the building and the entire site, allowing a consistent path from Clinton Avenue through the campus. Alleys and drives are now much like an Italian Piazza with raised stone banding and rough paving to slow down dangerous traffic.

The Pages

Like light thin sheets of paper, frosted glass walls with exposed, cantilevered edges float from the structure, flowing away from the building beyond. Native Arkansas sandstone defines the archive storage box and building plinth, signifying the weight of the materials stored within.

A soft story opens visually the level one galleries to three sides, allowing entrances to activate all sides of the building.

The glass planes fold and overlap, stepping back to protect the site's only mature tree, while evoking the imagery of pages.

A Place to Learn

The main research hall is an entire floor of the 1914 building, open much like the warehouse it once housed. The atrium functions as part of a high security plan, where librarians access the secured archive storage across the bridge.

History on Display

The entrance atrium blurs the edges between exterior and interior, building and site. The sandstone plaza bleeds into the atrium as the base of the archive storage box.

The patrons first images upon entering the building are of the galleries, image panels at elevator landings above, and the actual storage documents themselves.

A Shadow of History

History is a memory, an echo of the past…a shadow of what once was...but it is also fluid and active.

The Arkansas Studies Institute weaves history, research, pedestrians, and a restored streetscape together, healing a gaping wound in the urban fabric, while expanding environmental stewardship into the public realm and serving as a beacon of knowledge.

画廊赋予了街景更多活力，吸引着行人的目光
Galleries enliven the streetscape, drawing pedestrians in

Credits

Location: Arkansas, USA
Completion Year: 2009
Renovation Area: 4,550 m²
New Construction Area: 2,040 m²
Architect: Polk Stanley Wilcox Architects

The Rock 巨石
Wellington Airport International Passenger Terminal
惠灵顿国际机场客运航站楼

| Studio Pacific Architecture |

坚如岩石的外观融入独特的美学艺术，给人以视觉冲击
The rock façade is designed considering aesthetics, leaving a visual impact on people

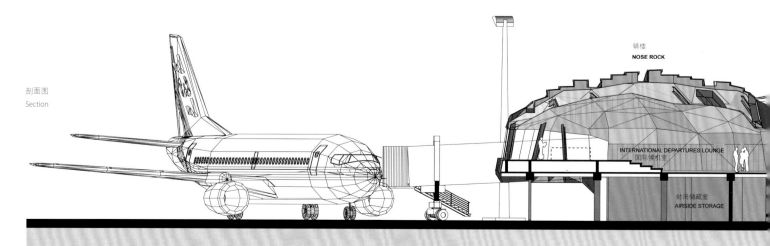

剖面图
Section

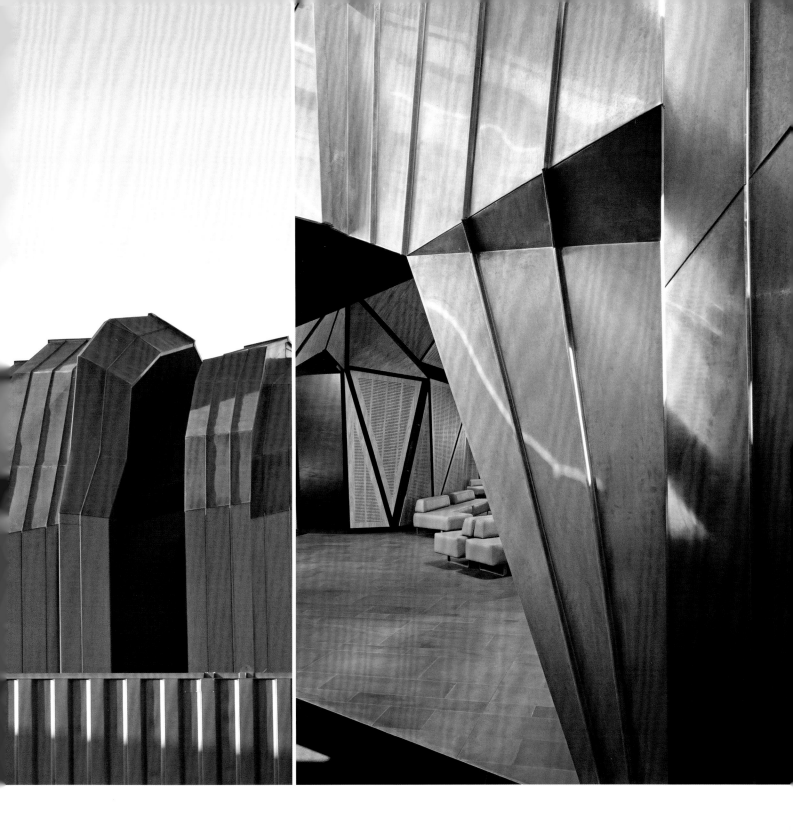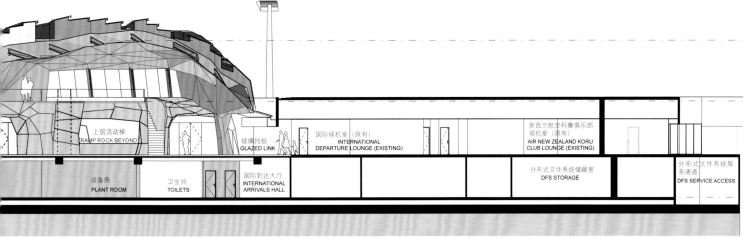

区位图
Site plan

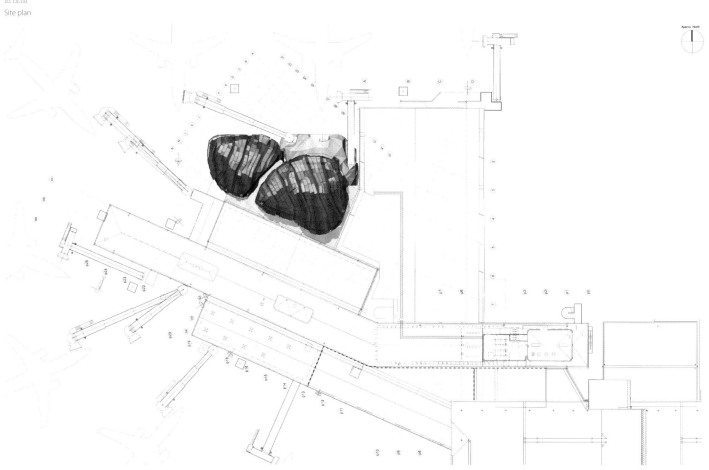

设计师想给每一位途经惠灵顿国际机场的游客留下美好记忆，因此航站楼采用了独特、犀利的美学设计，并将其融入周围宏大的氛围中。新建的机场航站楼外形如同一块坚硬、神秘的岩石，烘托着海浪呼啸的惠灵顿海岸。这里是世界各地航班的重要始发地，每天都浸盈在阳光和旅行的乐声中。"巨石"成为了这座岛屿上的一个定点，每天迎接着陆的航班，送走那些滑向海岸的乘客。航站楼内如同蚕茧般龟缩在崎岖封闭的外墙之内，既承接了机场外墙的多元性和厚重性，同时为游客提供了一个温暖、舒心的内部环境。

与大部分国际机场乏味无趣的内部装饰相比，"巨石"的内部结构透露着个性色彩。上方的大果镶板散发着柔和、温暖、甜蜜的光泽，与暗黑色钢板结构形成的冷漠形成强烈对比；充足的太阳光线照射进来，在这黑色柜台区域形成了光柱奇景。沿着屋顶裂缝照射进来的光线营造出航站楼的戏剧效果，同时来回走动的游客不经意间会注意到地板上的光圈——那全是由刮花玻璃表面映着灯光形成的。航站楼楼面分为数层，内部由道道缓坡连通。"巨石"的内部融合了许多经过重建的走廊区域平台，容纳的游客人数是以往的两倍，每小时接纳1 000名游客。

三角形的木制胶合板构成多面、波状的机场天花板，呼应着航站楼外立面的复杂几何形状。木制胶合板之间刻有凹槽，使会议空间空气流畅，同时极好地抑制了机场的混声回响。照明以及其他裸露在外的装置集中置于凹槽内，使内部线条整洁、流畅。而航站楼外部饱经风化的铜镀层表面包裹着楼内结构，连接内外空间，使人感觉这岩石需要被砸开，才能饱览机场内部的模样。

室内外对比
The dramatic contrast between soft interior and protective exterior

灯光下,内部的大果镶板散发着柔和与甜美的光泽
Macrocarpa panelling creates a soft honey glowing interior space under the light

剖面图
Section

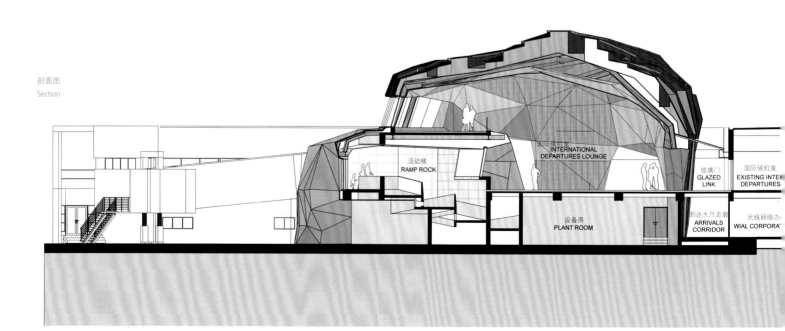

The brief for this airport terminal was to create a memorable visitor experience through a unique, edgy aesthetic that embodied a strong sense of place. The new terminal building was designed as a crusty, enigmatic rock that recalls the sea-battered Wellington coast. A radical departure from airports worldwide, preoccupied with imagery of lightness and flight, The Rock evokes the anchoring qualities of the land that rises to meet planes as they touch down and the coast that recedes as they depart. The interior was conceived as a cocoon-like space nestled inside this craggy protective exterior. It retains the organic qualities and rich materiality of the exterior while offering a warmer, more intimate interior environment.

In contrast to the bland interiors that typify most international airports, The Rock interior resounds with personality. Macrocarpapanelling creates a soft, warm honey-glow that contrasts dramatically with the dark-stained panels forming the negative spaces, with clustered pinpoints of light set like constellations into these black counter forms. A glazed fissure in the roof allows daylight into the space and creates a theatrical moment in the building, while small fractured glass apertures in the floor allow glimpses of arriving passengers below. Spaces unfold on varying levels and exploration is welcomed, with journeys through the interior gently modulated by a series of ramps. The interior of the Rock, along with a reconfiguration of the existing lounge areas, allows a doubling of lounge capacity to 1,000 passengers per hour.

Timber veneer paneling was triangulated to create a faceted, gently undulating ceiling that absorbs the building's complex exterior geometry. Slots carved into the timber panels allow air to pass through into a plenum space, which at the same time dampens the acoustic reverberation. Lighting and other exposed services are collected together in recessed service clusters to maintain the clean form of the interior linings. The weathered copper cladding of the exterior also wraps through into the interior, linking the interior and exterior and reinforcing the sense that the rock has been broken open to reveal the spaces inside.

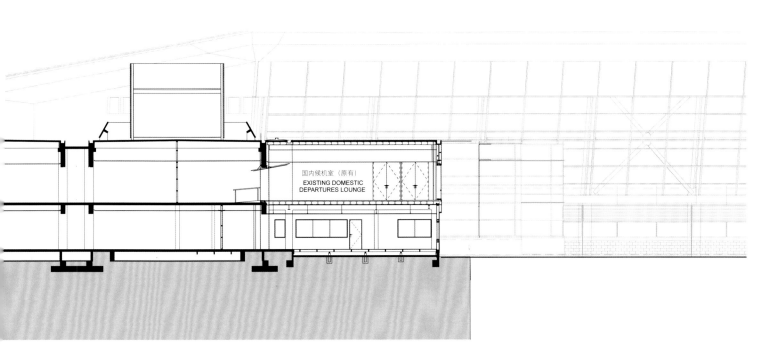

楼道
Rock stairs

Credits

Location: Stewart Duff Drive, Rongotai, Wellington, New Zealand
Completion Year: October 2010
Built Area: 2,100 m²
Floor Materials: Basalt Tiles
Walls and Ceiling: Macrocarpa Veneer Panels
Use: Airport International Passenger Terminal
Architects: Studio Pacific Architecture in association with Warren and Mahoney
Team: Nick Barratt-Boyes, Marcellus Lilley, Pamela Scott, Nick Acton-Adams, Ralph Roberts, Rodney Sampson, Claire Sharp
Client: Wellington International Airport Ltd
Constructor: Mainzeal

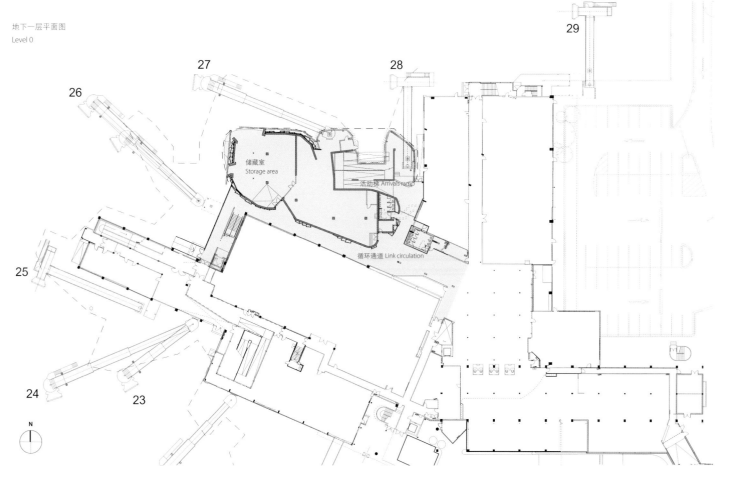

地下一层平面图
Level 0

户外平台上的桌椅与壁炉全由设计师精心打造,风格协调而独特,斟一杯红酒,放眼远处的葡萄园,此情此景,似乎融入了瓜达鲁普所有的美……
Exterior deck, furniture and fireplace designed by graciastudio, with the Valle de Guadalupe vineyards on the background

清新自然风
Endémico Resguardo Silvestre

瓜达鲁普阁楼式创意酒店

| graciastudio |

位于巴哈加利福尼亚的瓜达鲁普岛是墨西哥著名的酒乡,graciastudio在这里设计并建造了一座别样的酒店。这个酒店共有20套独立的房屋,每一套占地20平方米,由Grupo Habita饭店来经营。在这块占地99公顷的土地上,除了这个酒店,还有一个葡萄酒厂和一片居住区。

设计的一个原则是建筑不与土地直接接触,同时尽可能不干扰大自然环境。每一套房屋底部都用钢筋支撑,脱离地面,形成格局清晰的"生态阁楼"。耐候钢使房屋的颜色经久不衰,形成环境与建筑之间的和谐关系。房屋的设计灵感来自豪华宿营房的设想,不仅满足了宿营客人的基本需求,还形成建筑与自然环境之间的和谐而亲密的接触。

一栋栋独立的酒店客房在山坡上错落分布,成为眺望远处葡萄园与酿酒厂风景的绝佳地
Hillside view of frontal rooms facing towards the Valle de Guadalupe vineyards and wineries

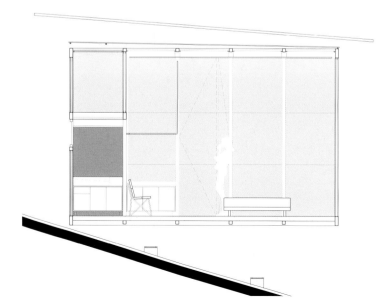

Located in Valle de Guadalupe "Mexico's Wine Country", Baja California, Endémico Resguardo Silvestre is a set of twenty independent rooms of twenty square meters each, operated by Grupo Habita, a Design Hotels member; established within a surface of 99 hectares, part of the Encuentro Guadalupe development, which includes a winery as well as a residential area.

One of the principal premises was not to interfere directly the land, as part of the philosophy of the project is to respect nature in every possible way. The availability of steel by our client lead to the design of the clean structure with this material, which elevates the skeleton of the room, named EcoLoft, to avoid contact with the soil. The employment of corten steel to cover it, which over time changes its color, achieving harmony between the environment and the building.

The approach of the design of the room comes from the concept of a "deluxe" camping house, covering the guest's basic needs, being in contact with nature and the environment.

酒店客房正立面景致
Hillside view of frontal rooms

酒店客房入口　　　　　　　　　客房内部景致
Entrance of the room　　　　　Inside view of the rooms

Hotel pool, rooms and Valle de Guadalupe vineyards on the background

Hillside view of room and deck

Credits

Location: Valle de Guadalupe, Ensenada, México
Year: 2011
Surface: 20 Rooms of 20 m² Each
Principal: Arq. Jorge Gracia
Collaborators: Javier Gracia, Jonathan Castellón, Braulio Lozano, Valeria Peraza
Construction: graciastudio
Photos: Luis Garcia

室内黑白色调织造简约经典风
Inside view of the black interior room

剧院临街一面
The theater from the main street

New Theater 蒙塔尔托迪卡斯特罗新剧院
in Montalto di Castro

| MDU ARCHITETTI |

纵剖面1
Longitudinal section A

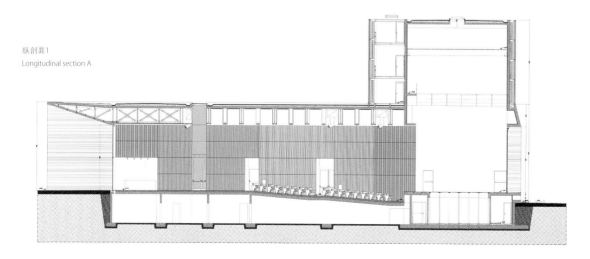

入口立面
The entrance facade

纵剖面2
Longitudinal section B

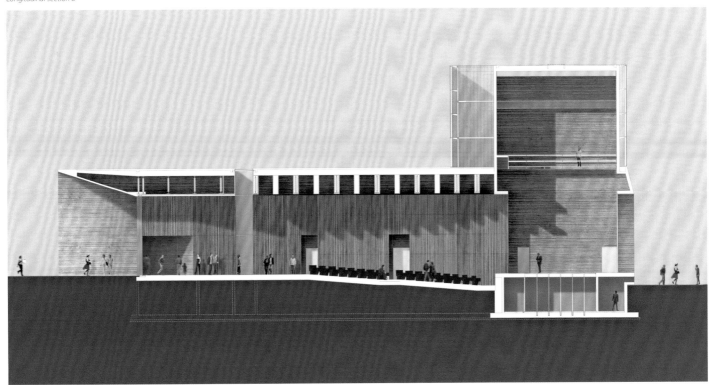

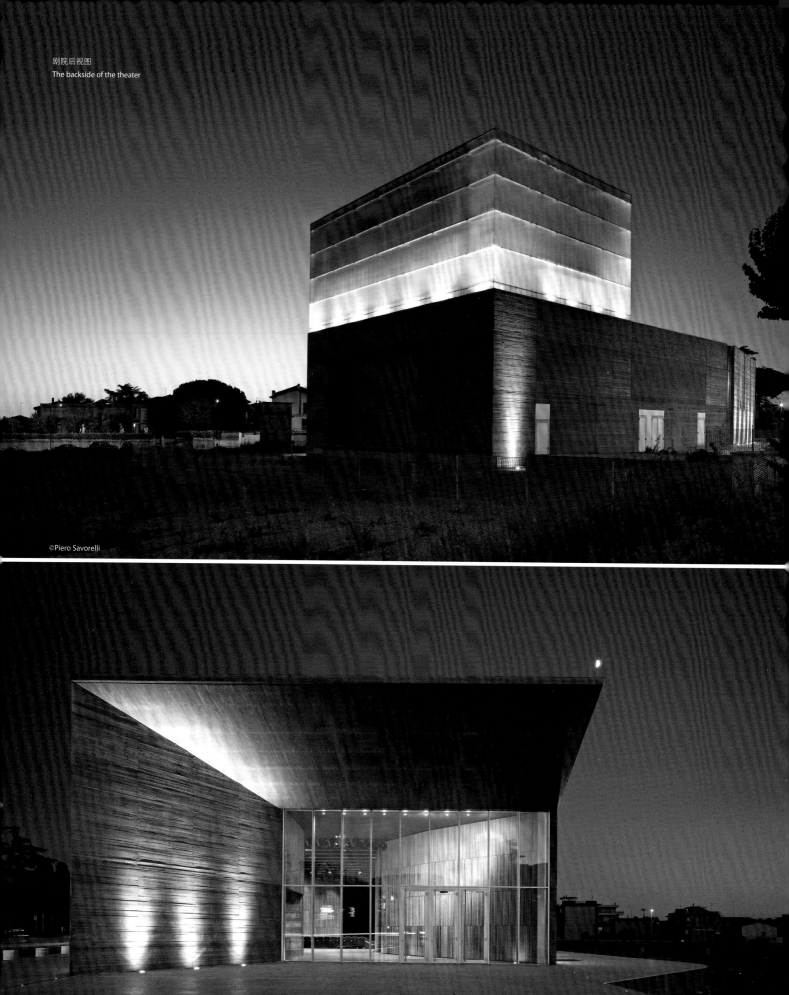

剧院后视图
The backside of the theater
©Piero Savorelli

剧院入口
The entrance to the theater
©Piero Savorelli

设计师旨在通过蒙塔尔托迪卡斯特罗新剧院的设计体现地域特色，同时还要通过建筑本身表现戏剧活动的魅力。

蒙塔尔托迪卡斯特罗起源于古代伊特鲁利亚文明，伊特鲁利亚人的遗迹表明，当时的建筑是由石灰华建成的立体几何房屋。如今，在世界人眼中，蒙塔尔托迪卡斯特罗是意大利最大的发电厂。剧院设计以"短接的时间线路"为设计理念，向蒙塔尔托迪卡斯特罗的文明发展致意，并通过独特的建筑元素达到这一目的：将古代伊特鲁利亚人与现代机器美学形成对比。新剧院如同由水泥构成的单块巨石，以微妙的颜色和质地变化为特色，剧院台塔显得十分轻盈。白天，由聚碳酸酯材料打造的凹槽结构沐浴在阳光下；夜间，剧院内的灯光使其演变成了一个巨大的"灯笼"，照亮了整个区域。

在通往历史中心的道路旁，设计师还用石灰华和水泥打造了座全新的露天广场，广场通往新剧院的入口，入口的悬挑屋檐格外引人注目。游客通过入口可以进入剧院，室内大堂和礼堂之间畅通无阻。木制的外墙上依稀看得见裂缝。这种厚重的形态通过多样化的建材得以弱化，由多种建材构成的立面将剧院围合起来，并引着观众来到被帘幕遮住的舞台处，等待帘幕神奇拉开的那一刻。

大礼堂内可容纳400人，而对应的户外场地有500个座位，在这里能更好地感受舞台的戏剧效果。

剧院大厅
The foyer of the theater
©Piero Savorelli

观众席、舞台及背景幕
The audience and the stage in the backdrop

立面与剖面详图
Elevation and section detail

从舞台看观众席
The audience as seen from the stage

平面图
Plan

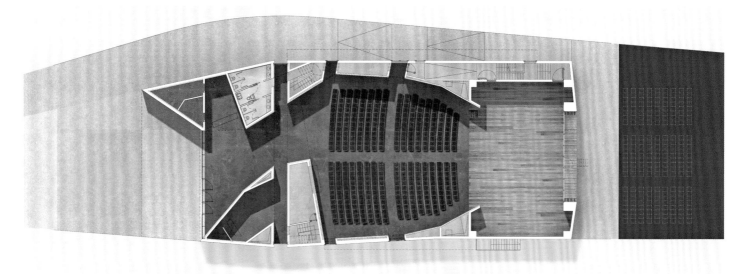

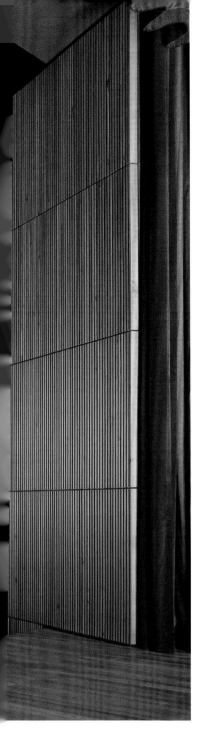

剧院入口
The entrance to the theater
©Piero Savorelli

剧院大厅
The foyer of the theater

The design for the New Theater in Montalto di Castro has a twofold objective: it is proposed as a conceptual model for measuring the territory and at the same time it attempts to express, through architecture, the magic of a theatrical event felt by the audience.

The territory of Montalto di Castro sinks its origins into Etruscan enthronization whose ruins attest to architecture comprised of large stereo metric masses in tufa; in the contemporary collective imagination Montalto di Castro evokes the world of the machines of the largest Italian power plant. The design proposes a temporal short circuit with respect to which the evolution of the territory is concentrated and expressed in a unique architectural moment: archaic Etruscan versus the aesthetics of the machine.

The new theater is a large concrete monolith characterized by subtle variations in color and texture, on which the fly tower appears to rest in an ethereal manner: an alveolar polycarbonate volume that dematerializes by day becoming indistinguishable from the sky, and lights up from within by night transforming into a large "lantern" on a territorial scale.

A new, extended piazza in travertine and concrete, designed as a diversion of the road providing access to the historic center, leads to the entrance of the New Theater identified by an impressive overhanging roof. It introduces visitors to a continuous environment in which the foyer and the auditorium flow freely into one another. The wooden walls, with their broken lines, create a space conceptually derived from the excavation of the concrete monolith. This morphological heaviness is contradicted by the vibration of the material that seems to envelope the space in a large curtain and introduces the spectator to the much awaited magical opening of the stage curtains.

The auditorium that seats 400 has its counterpart in the outdoor arena that seats 500, which can thus benefit from the theater stage.

大使馆为两座相对的L形建筑，通过庭院相连
The design of the building is based on two opposing L-shaped structures, which are linked by the garden

Swiss Embassy
瑞士大使馆

| PARAVANT ARCHITECTS |

位于喀麦隆雅温得的瑞士大使馆既融入了极强的公共属性和表现力，又在设计规划上着重考虑了社交隐私和安全，辩证地结合了二者，且具有地方特色。大使馆的复合结构采用了庭院的样式设计，迎合房间隐私和安全的需求，同时也蕴含了当地对空间、光照和气密性的独到理解。

大使馆被设计成两座相对的L形建筑，通过第三种元素连通：庭院。其中，有两座庭院空间作为领事馆官邸、大使馆官邸以及外交官员宿舍的主要设计元素，分布在三者之间。领事馆官邸是大使馆主要的公共核心区域，向南面朝主广场，以邻近广场入口的幕墙上标志性的瑞士国旗最为显眼。从标有瑞士国旗大楼二楼的外交官办公室处，可以望见周围的民舍和风景。大使馆官邸和外交官官邸之间隔着一条走廊，走廊两旁都是两

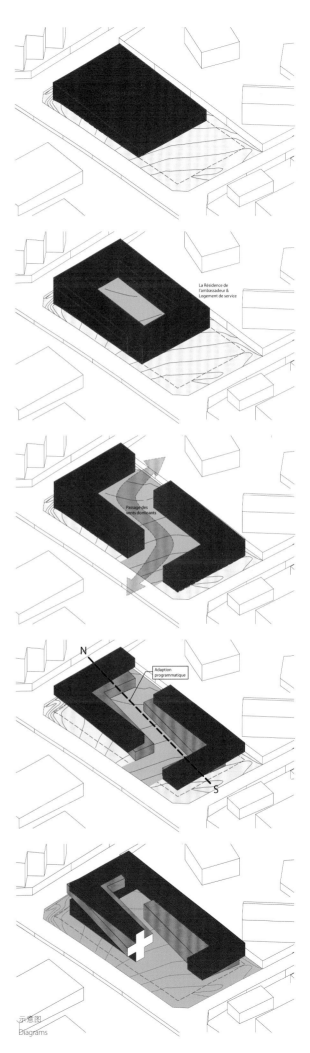

层高的建筑,那里有条通往底楼代表会大楼的限制性走道及前往居住区的私人通道,通道沿着庭院北部行进,在那里可以看见庭院外观和大使馆的全貌。二楼东侧的外交官员宿舍西面正对入口区,对面的庭院建于底楼远处,为住宿者开辟出一个室外平台,构成房间和外界之间的缓冲式景观。外交大使的私人办公套间位于二楼,和户外之间隔着一块穿孔金属墙,但是从那里依旧可以看到南面的庭院。瑞士大使馆底楼四周全是石墙,带有花纹的窗格既成为日常活动的私密屏风,又为室内引入自然风。内部庭院坐落着多层式花园,有助于交叉通风和纳凉,同时也提供了内部的聚会和活动空间。屏风与通风系统的结合设计,主要受非洲当地文化Mashrabiva建筑元素的影响,风格则趋向于传统的阿拉伯建筑。

示意图
Diagrams

内部庭院坐落着多层式花园，有助于交叉通风和纳凉，同时也提供了内部的聚会和活动空间
Within the walls are the courtyards with set of multi level gardens that help facilitate the cooling of the crossing breezes, while providing space for gatherings and events

剖面图1
Section A

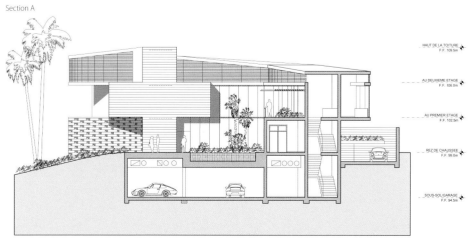

The Swiss Embassy designed for Yaoundé, Cameroun is a unique dialectic pairing of a highly public and representational building with extraordinarily sensitive requirements concerning privacy and security. The design of the compound uses the courtyard building type to address the need for privacy and security while also expressing the vernacular thoughts towards space, light and air that is typical in this region.

The design of the building is based on two opposing L-shaped structures, which are linked by a third element. This layout provides two courtyard spaces to serve the main program elements of the Chancellery, Ambassador Residence and the Staff Quarters. The Chancellery facilitating the main public and civic functions of the embassy faces south towards the main plaza and is marked by the iconic Swiss Cross shape of the façade near the entry gate of the plaza. From the Office of the Ambassador on the second floor of the Swiss Cross, there are views of the surrounding neighborhood and the distant landscape. The Ambassador Residence, connected to the Chancellery by a two-story corridor allowing controlled access to the representative functions on the ground floor and a private passage to the residence on the upper floor, follows the perimeter of the north courtyard, with views to the courtyard and of most of the embassy compound. The Staff Quarters on the east side of the second floor facing west towards the entry plaza and adjacent courtyard is rotated off plan from the first floor providing an outdoor terrace for their accommodations, and landscaped to provide a buffer between them and public sight. The private office suite of the Ambassador on the second floor is obscured from view of the public

剖面图2
Section B

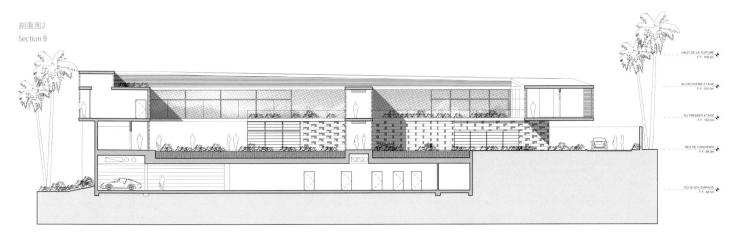

东侧立面
East elevation

西侧立面
West elevation

Credits

Architect: PARAVANT ARCHITECTS
Team: Halil Ramazan Dolan, Dipl.-Ing. Architekt, Nozomu Sugawara Architect, Toru Narita Architect, Kentaro Yamada Architect, Christian Kienapfel, Dipl.-Ing. Architekt, MarcusBrown, NCARB

by a perforated screen wall system, but still offers views out across the south courtyard toward the Swiss Cross. The site of the embassy is surrounded by a stonewall on the ground floor, composed of patterned openings, which serve as a privacy screen for the daily activities, but also acts as a medium for natural breezes and cross ventilation. Within the walls are the courtyards with set of multi level gardens that help facilitate the cooling of the crossing breezes, while providing space for gatherings and events. This approach to the joint function of the screen and ventilation system of the wall was influenced by the Mashrabiva elements that are a part of the local culture and anchored in traditional Arabic architecture.

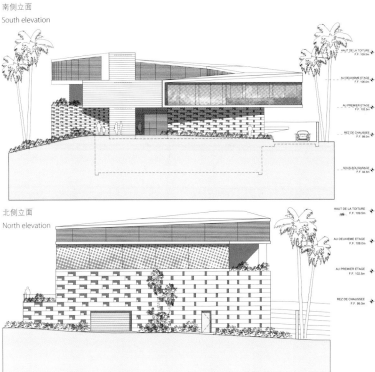

MiMA | MiMA豪华住宅区 |

| Rockwell Group |

规划背景

这座全新的MiMA豪华住宅是Related房地产公司斥资8亿美金打造的，约111 500平方米的60层高楼建筑的一部分，该建筑获得LEED绿色建筑银奖，它还包括纽约时报广场西侧的Yotel酒店和弗兰克·盖里设计的Signature剧院。MiMA豪华住宅因其坐落于曼哈顿中心而得名。住宅拥有663间出租单元房、151间分户出售单元房，分开的大厅，以及共享的设施和体育馆——为居住于此的青年置业人士提供他们所需的全部服务。

设计理念：出租区大厅和单元房

Rockwell Group设计事务所将大楼的出租区设计成一系列趣味丛生的空间。主大厅是一处社交平台，空间被精雕细琢的扭曲柚木屏风分割。这里的座位安排多种多样，鼓励这里的居民在任何时候都能驻足停留。大厅的雕塑风格延续至屋顶——一块不规则内凹天花板处悬挂着玻璃吊灯装置。房内另一处值得关注的地方是悬于邮件收发室上方的悬浮状玻璃球体和下垂吊灯。

这663间租赁单元房趣味丛生，值得玩味且极具现代风情。设计面向5大人群：年轻的单身女记者、年轻的单身男性网络工作者、喜欢旅行的著名男装设计师、好客的饭店业主和女演员夫妻家庭、以及爱好旅游、远足和美食的年长夫妻。

分户出售区大厅和单元房

分户出售的大厅比出租区大厅更有亲密的氛围，那里仅有151间单元房。定做的铜金属屏风、从顶部到地面的纹理玻璃、大理石地板以及定做的皮家具使得空间内部看上去复杂且高雅。

分户出售区大厅
Condo lobby

这151间分户出售的单元房面向3大人群：喜爱时尚的单身男性网络工作者、"博物馆控"和折中主义的夫妻，以及居住于康乃迪克州，但在曼哈顿需要一个住所，直至他们厌烦百老汇表演年长的夫妻。

附属设施楼层

位于三楼的附属设施楼层被分为2个区域——一个面朝MiMA豪华住宅的所有居民开放，另一个则仅对俱乐部会员居民开放。向所有人开放的地方名曰"狗仔城"，那里遍布供犬玩耍和美容的室内外区域，还有2间随时可以租赁的派对房。每间房内都拥有私密的平台，豪华的油漆面板、用厚实的皮料制成的真皮椅子，一处小厨房和吧台，以及先进的电视和音效系统。那里还有一处供烧烤的室外公共场所，以及可与房间一起租赁的一间备餐厨房。

3楼的另一边是仅对会员开放的俱乐部，那里到处可见扭曲的木制鳍状屏风，可以保证房内的私密性和交流，同时也保留了房间整体的通透和开放之感。在这动感无限的空间内会员们可以尽情地进行社交、放松或工作；休息室处有一座火炉，炉子由激光切割而成的铜材做成，装饰华

附属设施楼层
Amenities floor

美；而桌球房内的座椅全由奢华的皮料做成，还有任天堂游戏房以及拥有IMacs系列设备的商务中心，更有精致的食品售卖机与贴心的室内小厨房。另外俱乐部内还拥有一处放映室，屋顶呈曲折造型，由皮革面板制成，墙面被涂漆面板和紫色吸音泡沫材料包裹着，而小厨房内配备迷你冰箱，水槽以及爆米花机。健身俱乐部远远地坐落在尽头，昼营夜息，俱乐部内的游泳池四周被玻璃围合，地板铺设瓷砖，顶部呈波浪状外观，且小道遍布。桑拿区位于泳池最右侧，而每间更衣室内都配备蒸汽房。一部隶属健康俱乐部的电梯接送游客从3楼到达篮球城，篮球城位于下层地下室，电梯内缝制的布料象征着篮球的材料。在步入地下室之前就已经能感受到篮球的氛围。除了篮球馆和排球馆，还有两间房间用做拳击馆、瑜伽馆以及系统训练、健身中心。

游戏房
Game room

Background

The new MiMA luxury residence is part of Related Company's $800 million, 1.2 million square foot, 60-storied LEED-Silver complex, which also includes the Yotel Times Square West and a Frank Gehry designed Signature Theatre. MiMA is named for its central location in Midtown Manhattan. It is a LEED silver building with 663 rental units, 151 condo units, separate rental and condo lobbies, and a shared amenities floor and field house—providing all the services young professionals would ever need all in their place of residence.

Design Concept: Rental Lobby and Units

Rockwell Group designed a series of playful and fun "mixing" spaces for the rental portion of the building. The main lobby is a social "loft" with highly-detailed twisted teak screens that divide the spaces. This space offers a variety of seating arrangements that encourage residents to connect throughout the day. The sculptural nature of the lobby is continued on the ceiling, where amorphous shapes are carved out of the ceiling to make room for glass pendant light fixtures. Another focus of the space is the installation of suspended glass spheres and pendant lights that hang over the mail room.

The 663 rental units are fun, playful and modern, based on 5 personality types: single young female journalist, single young male athlete who works in IT, established male clothing designer with a penchant for traveling, an actress and

Lounge

restaurateur couple who love entertaining, and an older couple who love to travel and run and have a restaurant review blog.

Condo Lobby and Units
The condo lobby has much more of an intimate feeling than the rental lobby, as there are only 151 units. The sophisticated and elegant space is defined by custom bronze screens, floor-to-ceiling textured glass, marble floors, and custom leather furniture.

The 151 condo units are based on 3 personality types: a single young male who has a modern style and works in IT, a couple who frequent museums and have eclectic taste, and an older couple who live in Connecticut but need a place in Manhattan to stay since they see so many shows on Broadway.

Amenities Floor
The amenities floor on the third level is divided into 2 zones—one which is open to all MiMA residents, the other which is open only to residents who are members of the amenities club. Open to all is Dog City, an indoor/outdoor dog play area and grooming space, and 2 party rooms that are available for rent at

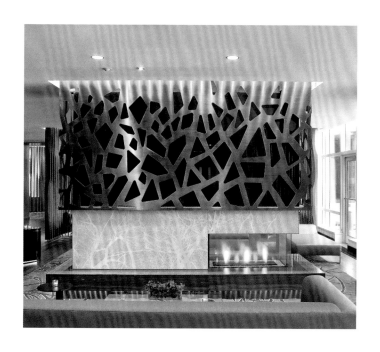

派对室
Party room

any time. The party rooms each have private terrace access, and boast lacquer paneled walls, an array of rich leather seating options, a kitchenette and bar, and a state-of-the-art TV and sound system. There is also a common outdoor space for barbecues and a catering kitchen available to rent along with the rooms. On the other side of the 3rd level is the member's only club, where twisted wood fin screens are used throughout the space to give privacy and intimacy to separate spaces, yet retain a fluid sense of openness throughout. The series of dynamic spaces where members can socialize, relax, or work are a lounge that features a fireplace with a laser cut bronze hood with decorative surround, a billiards room with luxurious leather seating, a Wii/game room, and a business center with a row of iMacs, food vending and a kitchenette. There is also a screening room with a billowing leather paneled ceiling, walls wrapped in lacquer panels and purple acoustic foam, and a kitchenette with a mini-fridge, sink and popcorn maker. At the end of the space is a health club operated by Equinox that boasts a glass-enclosed pool with a tile floor and ceiling with wave-like patterns and lanes demarcated. The sauna is right off the pool, and steam

邮件收发室
Mail room

Credits

Location: New York, NY, USA
Area: 4,088 m²
Completion Year: April 2011
Type: Condominium and Rental Residence
Services Provided: Architectural Design, Interior Design, Lighting Design, Custom Furniture and Fixture Design
Client: Related Companies

rooms are located in each locker room. There is a private health club elevator that travels from the 3rd floor to the Basketball City space in the sub-cellar field house—the leather stitching in the elevator mimics the materials of a basketball so that the experience begins before you step into the lower level. Besides courts for basketball or volleyball, there are also two more rooms for boxing, yoga, functional training and fitness.

北侧立面
North building elevation

Café 501 | 501咖啡馆 |

| Elliott + Associates Architects |

项目目标
客户要求：
1. 在午餐中提供新鲜、当地、快速、休闲的服务；
2. 夜晚提供不同的全方位服务；
3. 注重食物质量；
4. 能够吸引周边群众；
5. 招揽老顾客；
6. 氛围现代而舒适；
7. 以木头、砖、钢材和玻璃为主要建材；
8. 开放的厨房，温暖且友善；
9. 区别午餐与晚餐；

10. 保证新鲜；

11. 注重洗手间设计；

12. 私密性；

13. 服务品质；

14. 充满能量的空间；

15. 灵感的源泉；

建筑理念

手工制作

设计灵感——关键词

1. 空间的精神是什么？

   1) 有机

南侧过道门
Breezeway entry on south

宴会厅坐席细部
Banquet seating detail

带有剪影雕塑的中心餐厅
Center dining room looking north with silhouette sculpture

2) 朴实
3) 质感
4) 新鲜
5) 真实
6) 可信
7) 当地
8) 手工

2. 这种精神与你以何种方式联系？
1) 柔和
2) 植物状
3) 有生命力
4) 自然形态

3. 使用什么材料？
1) 编织
2) 天然
3) 木材
4) 陶土
5) 绳索
6) 石材
7) 皮革
8) 灯光
9) 气味

4. 怎样手工创造材料？
1) 编织
2) 砍伐
3) 鞣制
4) 拼合
5) 捆绑
6) 制造纹理
7) 雕刻
8) 堆砌
9) 折叠

5. 设计秘籍是什么？
1) 将食物和氛围交织在一起……
2) 食物和空间都是手工制作的……
3) 食物和空间都是自然的……
4) 食物和空间都是新鲜的……
5) 食物和空间都是真实的……
6) 混合的空间……
7) 交织的光线……

6. 两种情境
1) 白天开放的时候有大量的光线和自然氛围，以及一个开放的厨房；
2) 傍晚灯火通明、光影交错，适合私人聚会。

7. 引入地球、太阳、月亮、星星和星球的概念。

8. 它是一个私人空间……

从北侧中厅看餐厅
View into dining room from north outdoor patio

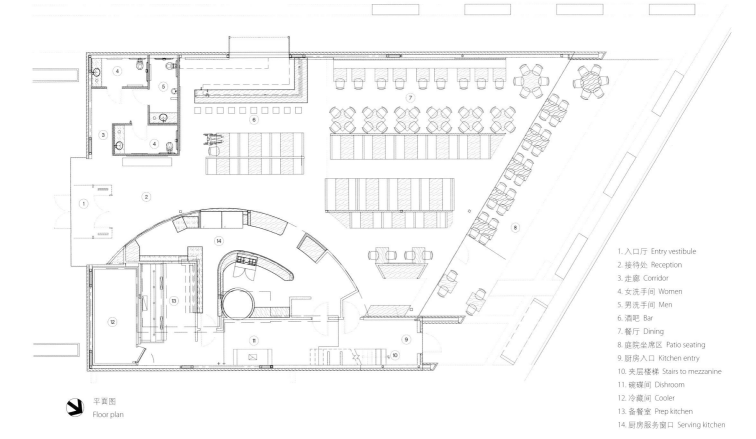

平面图
Floor plan

1. 入口厅 Entry vestibule
2. 接待处 Reception
3. 走廊 Corridor
4. 女洗手间 Women
5. 男洗手间 Men
6. 酒吧 Bar
7. 餐厅 Dining
8. 庭院坐席区 Patio seating
9. 厨房入口 Kitchen entry
10. 夹层楼梯 Stairs to mezzanine
11. 碗碟间 Dishroom
12. 冷藏间 Cooler
13. 备餐室 Prep kitchen
14. 厨房服务窗口 Serving kitchen

酒吧内景
Bar looking west with rotating ring

悬挂装饰木条细部
Hanging wood dowel detail

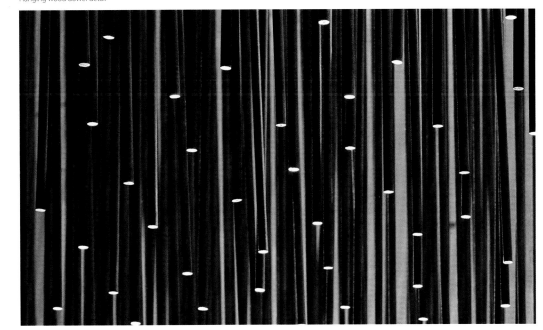

Project Goals

Key points from client discussion session:

1. Fresh, local, fast, casual service at lunch
2. Different at night—full service
3. Focus on quality food
4. The locals call us a neighborhood café.
5. We have lots of repeat business
6. More contemporary with a comfortable atmosphere....
7. Wood, brick, steel, glass
8. Open kitchen, warm and friendly
9. Lunch is 1st experience; Dinner is 2nd experience
10. In one word...fresh
11. "Character" in the toilets
12. Personal
13. Loyalty factor
14. Energy in space
15. Intuitive flow

Architectural Concept

Made by Hand

"Word Paintings" (Inspirations)

1. What is the spirit of the space?
   1). organic
   2). earthy
   3). textured
   4). fresh
   5). real
   6). authentic
   7). homegrown
   8). by hand
2. What forms connect you to this spirit?
   1). soft
   2). plant-like
   3). living forms
   4). form in its natural state
3. What materials will surround you?
   1). things woven
   2). things natural
   3). wood
   4). ceramics
   5). rope
   6). stone
   7). leather
   8). light
   9). aroma
4. How are materials created by-hand?

带旋转圆环的宴会厅
Banquet dining looking south towards bar with rotating ring

1). woven
2). hewn
3). tanned
4). knitted
5). tied
6). textured
7). carved
8). stacked
9). scored

5. What is the design recipe?
   1). The food and the atmosphere are woven together....
   2). The food and the space are handmade....
   3). The food and the space are natural....
   4). The food and the space are fresh....
   5). The food and the space are authentic....
   6). There is blended space....
   7). There is woven light....
   8). There are two moods
   9). Daytime...when it is open, light in weight, a natural atmosphere, and an open kitchen
   10). In the evening...there will be a warm glow, firelight, shadows and drama, it will be intimate
7. Bring in the earth, sun, moon, stars and planets....
8. It is a personal space....

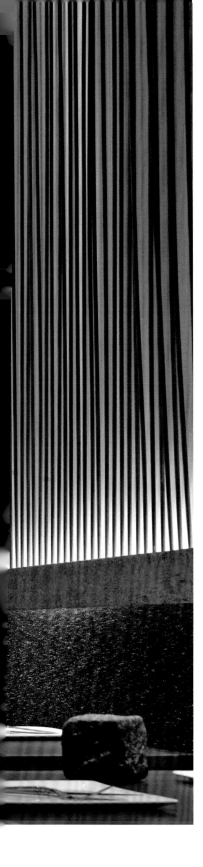

洗手间走廊
Toilet corridor

Credits

Location: Oklahoma City, USA
Completed Year: September 2010
Area: 501 m²
Architect: Elliott + Associates Architects
Project Team: Rand Elliott, David Ketch
Photographer: Scott McDonald, Hedrich Blessing
Client: Pete and Sheree Holloway
Contractor: Lingo Construction Services

Second Home | 第二家园 |

| Andre Kikoski Architect |

户外火炉
Open flame

放了烛台的树墩茶几
Stump tea tables with candles

第二家园的魅力使之成为当地的宠儿
Second Home's charisma has made it a local favorite

阿纳·雅各布森蛋椅
Arne Jacobsen Egg Chair

这个丹佛酒店餐厅的业主要求设计团队把这里打造成一个亚斯本或特鲁莱德式私人度假别墅，而不是传统的酒店餐厅。第二家园创造了一种家庭聚餐的感觉，以落基山脉和生活在那里的人们为灵感，营造出富丽堂皇、温馨热情的环境。第二家园将原本465平方米的空间分隔成一系列的空间，就像家里的房间：私人就餐区是书房，主餐区是开放的厨房/客厅，酒吧是娱乐室，半私人的就餐区是一个更正式的餐厅，天井是有白杨树和户外火炉的后院。

材料的选择突出第二家园是落基山脉的一部分。小马皮装饰家具和镶板幕墙、树皮纹理瓷砖、与石头产生对比的再生木，不锈钢和玻璃等丰富的纹理营造出迷人的空间。未加装饰的干砌石墙和粗糙的木板天花板构筑了素净的几何造型。阿纳·雅各布森蛋椅上的涂鸦和超凡的20世纪50年代意大利水晶灯等设计元素给空间带来一种诙谐和生气。超越单纯的美学范畴，第二家园源自于科罗拉多。所有的材料，从石头到木材，都取自于当地，当地的工匠在设计执行中扮演了整合的角色。第二家园轻松地将度假别墅的舒适愉悦融入现代美学和精致餐厅的智慧。这个空间将用餐者包围在温馨和谐中，使这座坐落于科罗拉多的建筑兼具了返璞归真之感和戏剧性。这种独特的体验吸引着一批又一批的回头客。第二家园的魅力使之成为当地的宠儿。它是人们真正想拥有的"第二家园"。

超凡的20世纪50年代意大利水晶灯等设计元素给空间带来一种诙谐和生气
Artisan-quality 1950s Italian chandeliers accent the space, bringing a stroke of whimsy and animation

材料的选择突出第二家园是落基山脉的一部分
The materials selected underpin the authenticity of Second Home as a part of the Rocky Mountains

Second Home transforms what was once one large 5,000 SF venue into series of spaces that feel like the rooms of a home

The Owner of this Denver hotel restaurant asked the design team to conjure up the feeling of someone's Aspen or Telluride vacation home, instead of a stereotypical hotel restaurant. The Second Home creates the feeling of being at someone's house for dinner in a rich, warm and welcoming environment by drawing inspiration from the character of the Rocky Mountains and the people who live there. Second Home transforms what was once one large 5,000 SF venue into series of spaces that feel like the rooms of a home: the private dining room is a study, the main dining room is an open kitchen / family room, the bar a recreational room, the semi-private dining room a more formal dining room, and the courtyard a backyard patio with the Aspen trees and open flame. The materials selected underpin the authenticity of Second Home as a part of the Rocky Mountains. Rich textures like pony-skin upholstery and paneled walls, bark tile, and reclaimed wood contrast with stone, stainless steel, and glass to create an intriguing and thoughtful material palette. The sober geometry of the plan is complimented by unembellished dry-stack stone walls and rough, wood-plank ceilings. Design elements such as graffiti-printed upholstery on Arne Jacobsen Egg Chairs and artisan-quality 1950s Italian chandeliers accent the space, bringing a stroke of whimsy and animation. More than simply aesthetics,

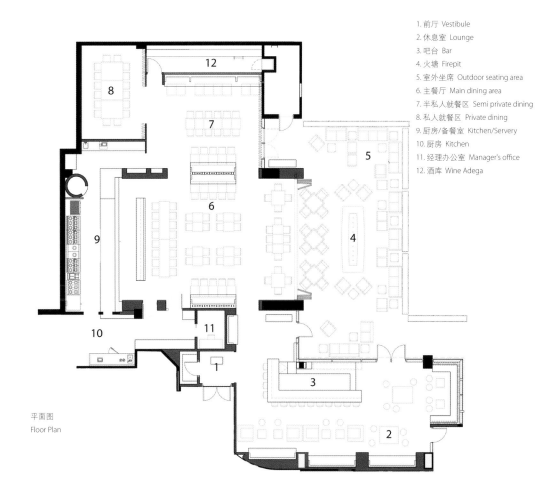

平面图
Floor Plan

1. 前厅 Vestibule
2. 休息室 Lounge
3. 吧台 Bar
4. 火塘 Firepit
5. 室外坐席 Outdoor seating area
6. 主餐厅 Main dining area
7. 半私人就餐区 Semi private dining
8. 私人就餐区 Private dining
9. 厨房/备餐室 Kitchen/Servery
10. 厨房 Kitchen
11. 经理办公室 Manager's office
12. 酒库 Wine Adega

Second Home is born of Colorado. All materials, from the stones to woods, were all locally sourced, and local craftsmen played an integral role in the execution of the design. Second Home effortlessly knits together the comforts and joys of a vacation home with the modern aesthetics and wit of a sophisticated restaurant. The space envelopes diners in warmth and hominess, and juxtaposes the austere and theatrical with architecture that grounds the restaurant in its Colorado setting. The result lends a distinct sense of place that draws back customers again and again. Second Home's charisma has made it a local favorite. It is a place people genuinely want to be; a Second Home.

Credits

Location: 150 Clayton Lane Denver, Colorado, USA
Completion Year: 2008
Area: 465m²
Project: Second Home Kitchen and Bar
Design Company: Andre Kikoski Architect
Photographer: Eric Laignel

未加装饰的干砌石墙
Unembellished dry-stack stone walls

树皮纹理瓷砖
Bark tile

办公室的前台如同一支箭,这是Yandex搜索引擎的非正式标志
There is a reception desk in form of an arrow in almost every Yandex office. This is an informal sign of the search engine—the brightest detail of neat website yandex.ru

Kazan Yandex Office

Yandex 喀山办公室

| za bor architects |

餐厅一角
Canteen fragment

这是俄罗斯最大的网络公司Yandex喀山的办公室，位于苏瓦尔广场商务中心的第16层。办公室面积647平方米，共有41个工作间。整个办公室布局呈梯形，其中线处是一条走廊，连接主入口和安全出口。

整个办公室空间沿走廊分布，走廊的两边各有四个开放式的办公区，每个办公区里能容纳6到13个工作间。另外还有一个独立的服务区，旁边是储藏间和IT管理员的办公室。行政办公室靠近入口，那里还设有前台。两间会议室、演讲厅和餐厅。往里走是员工办公室以及经理办公室，经理办公室外观类似多边形的玻璃灯，将两边的开放区域分隔开。必要的时候，经理办公室也可以通过窗帘与外界分隔开来。此处，磨砂玻璃还被用作了设计材料。走廊尽头是健身房和浴室。

为了让工作环境更舒适，办公室内的照明系统经过了精心设计，天花板上还安装了Ecophon吸音板——它具有极强的隔音作用。

Yandex的办公室最大的特点就是所有的办公间都是半开放式的，与走廊的开放空间分隔开来。在这间办公室内，设计师采用了圆形和工业化毛毯覆盖了工作间的墙的内表面和天花板，墙的外表面呈白色，而办公室天花板上的通信线呈同样的色调，与黑木地板形成显著对比。除了比较中性的黑白色调，办公室内的一些织物和家具还采用了陶土绿以形成对照，使办公室别有一番滋味。

主管办公室如同一只玻璃灯泡,通过有色玻璃与外隔绝,而窗帘更提升了办公室的私密性
Directors' chamber, organized as a glass light bulb. It is separated with tinted glass, but there are also black-out curtains for more privacy

Representative office of the largest Russian IT-company, Yandex in Kazan, is situated on the 16th floor of Suvar Plaza business center. With the area of 647m² it has only 41 working place. The office plan reminds a trapezium, while the midline of a trapezium is a corridor, leading from the main entrance to the fire exit.

The office volumes are organized in a corridor system—both side of the corridor has 4 open space blocks, designed for 6-13 working places. There is also a separate server zone, which borders with a storage and IT-administrator's office. Front-office volumes are located close to the main entrance—there is a reception desk, two meeting rooms, lecture-hall and canteen. Next there are working places and the chief's office, which looks like a glass polygonal lamp that functions as a zoning element, partially separating two nearest open spaces. If it is necessary, the chief's office may be isolated with curtains. Beyond that, matted glass panels are used as design element. At the end of the corridor there are gym and shower area.

To make the working areas more comfortable, the lighting estimations have been made, to correctly organize lighting, and the ceilings have been sheeted with Ecophon acoustic panels, which have sound-absorbing qualities.

Traditionally Yandex offices are famous for their informal rooms—unique semi-closed "cells", isolated from the corridor volume. In this office, the architects used rounded shapes and industrial carpet, which covers thoroughly the inside area of the two "cells". Outside they are white, as well as lines of communication up across the ceiling, which gives an effective contrast to black wooden floors. The only intensive color in the office, different from neutral monochrome colors, is a contrasting color scheme of terracotta-green, which was used repeatedly in textile and furnishings.

会议室的侧墙部分被有色玻璃覆盖，顶部开孔
This is what a sidewall of the "cell" meeting room looks like, but it is partially covered with tinted glass with a hole in the top

主管办公室
Inside directors' office

平面图
Plan

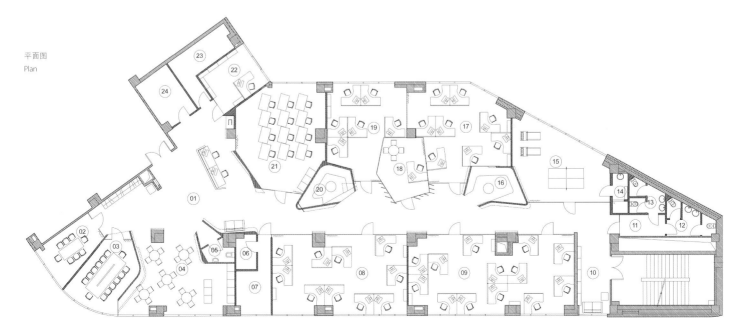

Yandex的每个办事处都设有用于非正式会议的交流区,位于喀山的办公室内设有两处交流区
Unique communicative "cells" zones for informal meetings are present in almost every Yandex office and there are two in Kazan' office